SAS® Guide to the REPORT Procedure: Usage and Reference

Version 6
First Edition

SAS Institute Inc.
SAS Campus Drive
Cary, NC 27513

The correct bibliographic citation for this manual is as follows: SAS Institute Inc., *SAS® Guide to the REPORT Procedure: Usage and Reference, Version 6, First Edition*, Cary, NC: SAS Institute Inc., 1990. 307 pp.

SAS® Guide to the REPORT Procedure: Usage and Reference, Version 6, First Edition

Contents

Reference Aids

Displays

Figures

Tables

Credits

Documentation

Composition	Ashlyn Bradshaw, Nancy Mitchell
Graphics	Creative Services Department
Proofreading	Carey H. Cox, Beth A. Heiney, Josephine P. Pope, Toni P. Sherrill, John M. West, Susan E. Willard
Technical Review	Johnny B. Andrews, Nancy Baker, Linda W. Binkley, Leslie B. Clinton, Brenda Hodge, Deborah J. Johnson, Christina A. Keene, Lynn P. Leone, Lynn H. Patrick, Bill Powers, Lisa M. Ripperton, Warren S. Sarle, Larry Stewart, Judy K. Theimer, Bobbie Wagoner, Amanda W. Womble
Writing and Editing	Rick Early, N. Elizabeth Malcom, Gary R. Meek, Helen F. Wolfson

Software

Alan R. Eaton developed the REPORT procedure. Program development includes designing, programming, debugging, and supporting the software as well as reviewing documentation.

Support

Development Testing	Johnny B. Andrews, Linda W. Binkley, Leslie B. Clinton, Helen F. Wolfson, Amanda W. Womble
Technical Support	Johnny B. Andrews, Christina A. Keene, Lynn H. Patrick
Quality Assurance	Deborah J. Johnson

Using This Book

Purpose

SAS Guide to the REPORT Procedure: Usage and Reference, Version 6, First Edition provides both usage and reference information for the REPORT procedure. The two types of information appear in separate parts of the book. The usage portion of the book presents a series of progressively more complicated tutorials that teach you how to use many features of PROC REPORT to create a variety of reports. It does not attempt to cover all the components of the REPORT procedure. The reference portion of the book provides complete descriptions of all REPORT commands and windows as well as a complete description of the REPORT language, but it does not attempt to teach you how to use the procedure.

"Using This Book" contains important information that will assist you as you use this book. It discusses the level of experience you need to use the REPORT procedure, describes the organization of the book, explains various conventions used throughout the book, and lists other books that may be useful to you.

Audience

SAS Guide to the REPORT Procedure is written both for users who have not extensively used the SAS System but who have some computer programming experience and know the fundamentals of programming logic and for experienced SAS users.

Prerequisites

The following table summarizes the SAS System concepts you need to understand in order to use *SAS Guide to the REPORT Procedure*.

You need to know how to	Refer to
invoke the SAS System at your site.	instructions provided by the SAS Software Consultant at your site.
allocate SAS data libraries and assign librefs. You should have an understanding of the concepts of SAS data libraries, SAS data sets, and SAS catalogs.	*SAS Language: Reference, Version 6, First Edition.*
use base SAS software. You need varying amounts of familiarity with the SAS System, depending on the level of complexity of the reports you want to produce. Because you can use DATA step programming in PROC REPORT, the more familiar you are with the DATA step, the more flexibility you have in producing reports. However, you can produce moderately sophisticated reports with very little knowledge of the DATA step.	*SAS Language: Reference, Version 6, First Edition*

How to Use This Book

This section describes how this book is organized and provides an overview of the information in it.

Organization

SAS Guide to the REPORT Procedure contains three parts. This section describes the purpose of each part and lists the chapters in each one.

Part 1: Usage

Part 1 begins with an overview showing you the various kinds of reports you can produce with the REPORT procedure. Five tutorials follow the overview. Each tutorial takes you through all the steps necessary to creating a particular report. Later tutorials rely on information presented in earlier ones. In addition, some tutorials use as a starting point a report developed in an earlier tutorial. Therefore, you should read these chapters in order.

Chapter 1, "An Overview of the REPORT Procedure"

Chapter 2, "Creating a Simple Report"

Chapter 3, "Enhancing Your First Report"

Chapter 4, "Grouping and Summarizing Data"

Chapter 5, "Using Some Advanced Features of the REPORT Procedure"

Chapter 6, "Using the COMPUTE Window"

Part 2: Reference

Part 2 begins with a description of how to use the REPORT procedure in a windowing environment. Chapters detailing the commands and windows available in the windowing environment follow. The last chapter describes the details of the REPORT language. You use the language either to create a report in a nonwindowing environment or to create a partial report that you can then modify in a windowing environment.

Chapter 7, "Using the REPORT Procedure in a Windowing Environment"

Chapter 8, "REPORT Commands"

Chapter 9, "REPORT Windows"

Chapter 10, "The REPORT Language"

Part 3: Appendix

The appendix contains the DATA step that produces the data set that all the reports in this book use.

What You Should Read

All users should begin by reading Part 1. If you can work in a windowing environment, you will probably master the material more easily if you actually work through the tutorials, creating the reports yourself. When you have gained some familiarity with the procedure, refer to the reference section for details. In Part 2, read Chapter 7 first. You can read the other chapters in any order. In fact, you can read isolated portions of chapters in the reference section to get the information you need.

If you are using the procedure in a nonwindowing environment, reading Part 1 is still useful because it provides you with an understanding of the way the REPORT procedure structures reports and of the terminology used to discuss reports. After reading the usage material, refer to Chapter 10, which describes the REPORT language. The examples in this chapter show how to use the REPORT language to create the same reports the tutorials create.

Reference Aids

SAS Guide to the REPORT Procedure is organized to make information easy to find. The following features are provided for your easy reference:

Contents   lists part and chapter titles, major subheadings, and the first page of each.

Reference Aids   lists all displays, figures, and tables in this book.

chapter tables of contents	list major and minor subheadings within a chapter and the first page of each.
Glossary	defines terms related to the REPORT procedure and report-writing.
Index	provides the page numbers where specific topics, concepts, statements, and options are discussed.

Chapters 8, 9, and 10 provide detailed reference information for using the REPORT procedure in a windowing and nonwindowing environment. The structure of each of these chapters generally follows a standard format.

In Chapter 8 the discussion of each REPORT command includes the following sections:

header	names the command and summarizes its function.
Action Bar	identifies a selection in the action bar in the REPORT window. When you make this selection, you access the pull-down menu where the command appears.
Description	describes the function of the command.
Usage	describes how to use the command.
Illustration	shows the visual affect of the command on the report.
See Also	provides references for related information.

In Chapter 9 the discussion of each REPORT window includes the following sections:

header	names the window and summarizes its function.
Display	shows a picture of the window.
Access	describes how to open the window.
Description	describes the function of the window. This section includes information on each choice in the window.

Chapter 10 describes the use of the REPORT language. The introduction to this chapter provides the high-level syntax for the REPORT procedure. A description of each statement follows. The statements are documented in alphabetic order except for the PROC REPORT statement, which appears first. The description of a statement includes the following sections:

header	names the statement and summarizes its function.
Syntax	provides the detailed syntax for the statement.
Requirements	describes any required arguments.
Options	describes optional arguments. Options appear in alphabetic order.

CONVENTIONS

This section explains the various conventions used in presenting text, SAS language syntax, file and library references, examples, and printed output in this book. The following terms are used in discussing syntax:

keyword   is a literal that is a primary part of the SAS language. A literal must be spelled exactly as shown. Keywords in this book are statement names.

argument   is an element that follows a keyword. It is either literal or user-supplied. It has a built-in value or has a value assigned to it.

                    Arguments that you must use are *required arguments*. Other arguments are *optional arguments*, or more simply, *options*.

value   is an element that follows an equals sign. It assigns a value to an argument. It may be a literal or a user-supplied value.

Note: The SAS System is case-sensitive only to characters enclosed in quotation marks. Thus, even though typographical and syntax conventions distinguish between uppercase and lowercase for clarity, you can generally enter SAS statements in uppercase or lowercase. When you are entering text in quotation marks, use case exactly as you want it to appear in your report.

Typographical Conventions

You will see several type styles in this book. The following list explains the meaning of each style:

roman   is the standard type style used for most text in this book.

UPPERCASE ROMAN   is used for SAS statements, variable names, and other SAS language elements when they appear in the text.

italic   defines new terms and emphasizes important information.

`monospace`   is used for examples of SAS statements and for references in text to the values of character variables. Monospace is also used in text to refer directly to any value in a window that you type in or can type over.

bold   is used in text to refer directly to any part of a window you can select and to instructions and field names that appear in a window. Typically, these are parts of the window you cannot change.

Note: Some windows contain things that you can both type over and select. For example, in the REPORT window, you can select a column header to perform a task or you can reword the header by typing over it. In such cases, the use of bold or monospace is determined by the context of the discussion.

Syntax Conventions

Type styles have special meanings when used in the presentation of syntax in this book. The following list explains the style conventions for the syntax sections. Remember that except for text in quotation marks, the SAS System is not case-sensitive.

UPPERCASE BOLD   identifies SAS keywords, which in this book are statement names (for example, **PROC REPORT** or **DEFINE**). You must spell the keyword exactly as it appears in the syntax statement.

UPPERCASE ROMAN   identifies arguments and values that are literals (for example, BEFORE and AFTER). You must use the exact spelling that appears in the syntax statement.

italic   identifies arguments or values that you supply. Items in italic can represent user-supplied values that are either

□   nonliteral values assigned to an argument (for example, *number-of-panels* in PANELS=*number-of-panels*)

□   nonliteral arguments (for example, *weight-variable* in WEIGHT *weight-variable*).

In addition, an item in italics can be the generic name for arguments from which you can choose (for example, *attribute-list* or *usage*).
Note: The use of the suffix *list* implies that you can specify more than one attribute, whereas the lack of the suffix implies that you can specify only one usage.

The following symbols may appear in syntax statements.

< > (angle brackets)   identify optional arguments. Any argument not enclosed in angle brackets is required.

| (vertical bar)   indicates that you can choose one element from a group. Values separated by bars are mutually exclusive.

. . . (ellipsis)   indicates that you can repeat the argument or group of arguments following the ellipsis any number of times. If the ellipsis and the following argument are enclosed in angle brackets, they are optional.

The following examples illustrate these syntax conventions.

BREAK BEFORE | AFTER *break-variable* < /*break-option-list*>;

□ **BREAK** is in uppercase bold because it is a SAS keyword, the name of a statement. You must spell it as shown. The remaining elements in this statement are required arguments and options.

□ BEFORE | AFTER is not enclosed in angle brackets because it is required. The vertical bar indicates that you can use either BEFORE or AFTER but not both. The arguments are in uppercase to indicate that they are literals and that you must spell them as shown.

□ *break-variable* is not enclosed in angle brackets because it is a required argument. It is in italic because it is a value you must supply.

□ *break-option-list* is enclosed in angle brackets because it is an optional argument. It represents a list of values from which you can choose one or more elements. If you use any options, a forward slash (/) must precede the first one.

□ The ending semicolon (;) is not enclosed in angle brackets because it is required.

DEFINE *item*
    </ <usage>
      <attribute-list>
      <option-list>
      <justification>
      <'column-header-1'<. . . 'column-header-n'>>
      *<*COLOR=*color>>*;

□ **DEFINE** is in uppercase bold because it is a SAS keyword, the name of a statement. You must spell it as shown. The remaining elements in this statement are arguments and values for arguments.

□ *item* is in italic because you supply the name of any item in the report. It is not enclosed in angle brackets because it is required.

□ *usage* represents a list of options from which you can choose only one option. *Usage* is enclosed in angle brackets because it is optional.

□ *attribute-list* and *option-list* represent lists of options from which you can choose one or more options. Both are enclosed in angle brackets because they are optional.

□ *justification* represents a list of options from which you can choose only one option. *Justification* is enclosed in angle brackets because it is optional.

□ *column-header-1* is enclosed in angle brackets because it is an optional argument. The words are in italic because they are values you must supply. In this case, *column-header-1* is the text for the first line of the header for the column containing *item*.

□ *column-header-n* is enclosed in angle brackets because it is an optional argument. The words are in italic because they are values you must supply. In this case, *column-header-n* is the text for the *n*th line of the header for the column containing *item*. The ellipsis indicates that you can use as many column headers as you want.

Note: If you use any options, a forward slash (/) must precede the first one.

Conventions for Displays, Output, and Examples

The displays and output in this book were created using the NODATE option. The line size used in most reports in displays is 78; the page size is 20. In most cases, the line size used in the reports in output is 78; the page size is 28. In some cases, a page size of 60 is used to fit a long report on a single page of output. Your displays and output may differ from those shown in the book if you use different options or different values for your line size and page size.

    Note: The default values for line size and page size vary from one operating system to another.

    The explanations of some pieces of output include a numbered list. These numbers correspond to the boldface numbers in the example output.

    Currently, colors you specify for a report appear only when PROC REPORT displays the report in the REPORT window. Thus, output in this book is black and white, whereas displays use color. Colors in displays are not the default colors. Therefore, the colors in your displays will differ from the colors presented in this book.

    The examples in this book use the LIBNAME statement to associate a logical reference (libref) with a SAS data library. On some operating systems you can use operating system control language to make these associations. On the VSE operating system, you must use operating system control language.

Additional Documentation

SAS Institute provides many publications about software products of the SAS System and how to use them on specific operating systems. For a complete list of SAS publications, you should refer to the current *Publications Catalog*. The catalog is produced twice a year. You can order a free copy of the catalog by writing to the following address:

> SAS Institute Inc.
> Book Sales Department
> SAS Campus Drive
> Cary, NC 27513

    The following is a list of selected SAS publications that may be helpful to you as you learn to use the REPORT procedure:

☐ *SAS Language and Procedures: Introduction, Version 6, First Edition* (order #A56074) provides information for users who are unfamiliar with the SAS System or any other programming language.

☐ *SAS Language: Reference, Version 6, First Edition* (order #A56076) provides detailed reference information about all elements of the SAS language except procedures. While you are using PROC REPORT, you may find the chapters on statements, formats, functions, and the SAS Text Editor particularly useful.

☐ *SAS Procedures Guide, Version 6, Third Edition* (order #A56080) provides detailed reference information about other procedures in base SAS software. In particular, you may find the chapters on the CATALOG, DATASETS, and FORMAT procedures useful.

□   *SAS Guide to Macro Processing, Version 6, Second Edition* (order #A56041) provides a tool for further customizing your reports.

□   SAS documentation for your host operating system provides information about the operating-system-specific features of the SAS System for your operating system.

Part 1

Usage

Chapter **1** An Overview of the REPORT Procedure

Introduction

The REPORT procedure combines features from the PRINT, MEANS, and TABULATE procedures with features of DATA step report writing into a powerful report-writing tool. You can use the REPORT procedure in one of three ways:

☐ in a windowing environment with a prompting facility that guides you as you build a report.

☐ in a windowing environment without the prompting facility.

☐ in a nonwindowing environment. In this case you submit a series of statements with the PROC REPORT statement, just as you do in other SAS procedures.

You can store a report definition and use it in either a windowing or nonwindowing environment to create the same report for any SAS data set that contains the variables used in the report definition.

This chapter shows you a range of reports that you can create with the REPORT procedure and briefly discusses some of the features of these reports. The chapter also illustrates some of the terminology this book uses to discuss PROC REPORT and the reports you can create with it. For more detailed definitions, see the glossary at the end of this book.

The tutorials that follow in the next five chapters show you how to create all but the simplest of the reports in this chapter. (This chapter shows you how to create the two simplest reports. The tutorials lead you through the step-by-step processes you use to create the reports in a windowing environment. At the end of each tutorial is a listing of the REPORT language you can use to create the same report in a nonwindowing environment. Chapter 10, "The REPORT Language," explains these listings in detail.

Sample Reports

This section presents eight reports created by the REPORT procedure. The reports range from extremely simple to fairly complex. The text that accompanies each report gives a brief explanation of the features used to create the report. All reports use a permanent SAS data set named REPORT.BUDGET. This data set contains two quarters of financial data for a company that makes video tapes. The variables in the data set are QTR (the quarter), DEPT (a department in the company), ACCOUNT (an account in a department), BUDGET (the amount budgeted for an account), and ACTUAL (the amount spent by an account).

The first two reports illustrate the two types of reports you can get if you invoke the REPORT procedure without any options or subordinate statements.

Boldface numbers in a report correspond to the numbers in the list that follows the output.

Producing the Simplest Reports

If you run the REPORT procedure on the data set REPORT.BUDGET, which contains both character and numeric variables, without using any options or subordinate statements, the report it creates is similar to the output you would get from running a comparable PROC PRINT step. The following SAS statements use the nonwindowing environment to create the report in Output 1.1:

```
libname report 'SAS-data-library';

proc report data=report.budget;
run;
```

Output 1.1
Producing the
Default Report for
a Data Set
Containing Both
Character and
Numeric Variables

```
                              The SAS System    1                         1
                     2
              QTR  DEPT        ACCOUNT       BUDGET       ACTUAL
               1  Staff       fulltime  $130,000.00  $127,642.68
               2  Staff       fulltime  $165,000.00  $166,345.75
               1  Staff       parttime   $40,000.00   $43,850.12
               2  Staff       parttime   $60,000.00   $56,018.96
         3     1  Equipment   lease      $40,000.00   $40,000.00
               2  Equipment   lease      $40,000.00   $40,000.00
               1  Equipment   purchase   $40,000.00   $48,282.38
               2  Equipment   purchase   $20,000.00   $17,769.15
               1  Equipment   tape        $8,000.00    $6,829.42
               2  Equipment   tape       $12,000.00   $11,426.73
               1  Equipment   sets        $7,500.00    $8,342.68
               2  Equipment   sets        $7,500.00    $8,079.62
               1  Equipment   maint      $10,000.00    $7,542.13
               2  Equipment   maint      $12,000.00   $10,675.29
               1  Equipment   rental      $4,000.00    $3,998.87
               2  Equipment   rental      $6,000.00    $5,482.94
               1  Facilities  rent       $24,000.00   $24,000.00
               2  Facilities  rent       $24,000.00   $24,000.00
               1  Facilities  utils       $5,000.00    $4,223.29
               2  Facilities  utils       $3,500.00    $3,444.81
               1  Facilities  supplies    $2,750.00    $2,216.55
               2  Facilities  supplies    $2,750.00    $2,742.48
               1  Travel      leases      $3,500.00    $3,045.15
               2  Travel      leases      $4,500.00    $3,889.65
               1  Travel      gas           $800.00      $537.26
               2  Travel      gas         $1,200.00      $984.93
               1  Other       advert     $30,000.00   $32,476.98
               2  Other       advert     $30,000.00   $37,325.64
               1  Other       talent     $13,500.00   $12,986.73
               2  Other       talent     $19,500.00   $18,424.64
               1  Other       musicfee    $3,000.00    $2,550.50
               2  Other       musicfee    $5,000.00    $4,875.95
```

This report illustrates the following features:

1. a title.

2. a column for each variable in the data set, including a column header containing the name of the variable (or label if the data set contains a label for the variable).

3. a row of data for each observation in the data set. The rows appear in the same order as the observations occur in the data set. Each of these rows is a *detail row*.

If you run the REPORT procedure without any options or subordinate statements on a data set that contains only numeric variables, PROC REPORT produces a one-line summary report that shows the sum of each variable over all observations in the data set. For instance, if the data set contained only the variables BUDGET and ACTUAL, PROC REPORT would produce the report in Output 1.2.

Output 1.2
Producing the
Default Report for
a Data Set with
Only Numeric
Variables

```
                          The SAS System                          1

                        BUDGET      ACTUAL
                    $775,000.00  $780,011.28
```

Ordering Rows

The report in Output 1.3 is somewhat more sophisticated than the default report.

Output 1.3
Ordering the Rows
of a Report

```
                                        The SAS System                          1
                                                            1
                                                 Amount      Amount
                  Quarter  Department  Account   Budgeted     Spent
                       1   Equipment   lease    $40,000.00  $40,000.00
      2                                maint    $10,000.00   $7,542.13
                                       purchase $40,000.00  $48,282.38
                                       rental    $4,000.00   $3,998.87
                                       sets      $7,500.00   $8,342.68
                                       tape      $8,000.00   $6,829.42
                           Facilities  rent     $24,000.00  $24,000.00
                                       supplies  $2,750.00   $2,216.55
                                       utils     $5,000.00   $4,223.29
                           Other       advert   $30,000.00  $32,476.98
                                       musicfee  $3,000.00   $2,550.50
                                       talent   $13,500.00  $12,986.73
                           Staff       fulltime $130,000.00 $127,642.68
                                       parttime $40,000.00  $43,850.12
                           Travel      gas         $800.00     $537.26
                                       leases    $3,500.00   $3,045.15
                       2   Equipment   lease    $40,000.00  $40,000.00
                                       maint    $12,000.00  $10,675.29
                                       purchase $20,000.00  $17,769.15
                                       rental    $6,000.00   $5,482.94
                                       sets      $7,500.00   $8,079.62
                                       tape     $12,000.00  $11,426.73
                           Facilities  rent     $24,000.00  $24,000.00
                                       supplies  $2,750.00   $2,742.48
                                       utils     $3,500.00   $3,444.81
                           Other       advert   $30,000.00  $37,325.64
                                       musicfee  $5,000.00   $4,875.95
                                       talent   $19,500.00  $18,424.64
                           Staff       fulltime $165,000.00 $166,345.75
                                       parttime $60,000.00  $56,018.96
                           Travel      gas       $1,200.00     $984.93
                                       leases    $4,500.00   $3,889.65
```

The differences between this report and the default report are as follows:

1. The column headers no longer contain uppercase variable names. Instead, the headers use mixed case and are more informative than the variable names themselves. The headers are not labels stored in the data set; rather, they are for use only in the report.

2. The REPORT procedure orders the rows first by their values of 'Quarter', next by their values of 'Department' within 'Quarter', and finally, by their values of 'Account' within 'Department'.

Computing a New Variable and Adding Further Enhancements

The report in Output 1.4 uses several more features of the REPORT procedure to further enhance the report in Output 1.3.

Output 1.4
Enhancing the
Ordered Report

```
                          1   Report of Departments                        1
                              by Quarter

                                    Amount        Amount        3
  2  Quarter  Department  Account   Budgeted      Spent       Balance
     --------------------------------------------------------------------

       1  Equipment   lease      $40,000.00    $40,000.00        $0.00
                      maint      $10,000.00     $7,542.13    $2,457.87
                      purchase   $40,000.00    $48,282.38   $-8,282.38
                      rental      $4,000.00     $3,998.87        $1.13
                      sets        $7,500.00     $8,342.68     $-842.68
                      tape        $8,000.00     $6,829.42    $1,170.58
          Facilities  rent       $24,000.00    $24,000.00        $0.00
                      supplies    $2,750.00     $2,216.55      $533.45
                      utils       $5,000.00     $4,223.29      $776.71
          Other       advert     $30,000.00    $32,476.98   $-2,476.98
                      musicfee    $3,000.00     $2,550.50      $449.50
                      talent     $13,500.00    $12,986.73      $513.27
          Staff       fulltime  $130,000.00   $127,642.68    $2,357.32
                      parttime   $40,000.00    $43,850.12   $-3,850.12
          Travel      gas           $800.00       $537.26      $262.74
                      leases      $3,500.00     $3,045.15      $454.85
```

```
                              Report of Departments           4   2
                              by Quarter

                                    Amount        Amount
     Quarter  Department  Account   Budgeted      Spent       Balance
     --------------------------------------------------------------------

       2  Equipment   lease      $40,000.00    $40,000.00        $0.00
                      maint      $12,000.00    $10,675.29    $1,324.71
                      purchase   $20,000.00    $17,769.15    $2,230.85
                      rental      $6,000.00     $5,482.94      $517.06
                      sets        $7,500.00     $8,079.62     $-579.62
                      tape       $12,000.00    $11,426.73      $573.27
          Facilities  rent       $24,000.00    $24,000.00        $0.00
                      supplies    $2,750.00     $2,742.48        $7.52
                      utils       $3,500.00     $3,444.81       $55.19
          Other       advert     $30,000.00    $37,325.64   $-7,325.64
                      musicfee    $5,000.00     $4,875.95      $124.05
                      talent     $19,500.00    $18,424.64    $1,075.36
          Staff       fulltime  $165,000.00   $166,345.75   $-1,345.75
                      parttime   $60,000.00    $56,018.96    $3,981.04
          Travel      gas         $1,200.00       $984.93      $215.07
                      leases      $4,500.00     $3,889.65      $610.35
```

This report contains the following new features:

1. a customized title

2. a visual break, consisting of underlining and a blank line, between the column headers and the first detail row on each page

3. a new variable, BALANCE, calculated from columns in the report, under the column header 'Balance'

4. a page break when the value of 'Quarter' changes.

Grouping and Summarizing Observations

The report in Output 1.5 groups and summarizes the observations in the data set.

Output 1.5
Grouping and
Summarizing
Observations

```
                                                                        1

                          Year-to-Date Financial Status
                             Grouped by Department

                                 Amount        Amount
          Department  Account   Budgeted        Spent        Balance
          Equipment   lease    $80,000.00    $80,000.00          $0.00
                      maint    $22,000.00    $18,217.42      $3,782.58
   1                  purchase $60,000.00    $66,051.53     $-6,051.53
                      rental   $10,000.00     $9,481.81        $518.19
                      sets     $15,000.00    $16,422.30     $-1,422.30
                      tape     $20,000.00    $18,256.15      $1,743.85
                        3      -----------   -----------    -----------
                             3 $207,000.00   $208,429.21    $-1,429.21  2
                        3      -----------   -----------    -----------

          Facilities  rent     $48,000.00    $48,000.00          $0.00
                      supplies  $5,500.00     $4,959.03        $540.97
                      utils     $8,500.00     $7,668.10        $831.90
                               -----------   -----------    -----------
                               $62,000.00    $60,627.13      $1,372.87
                               -----------   -----------    -----------

          Other       advert   $60,000.00    $69,802.62     $-9,802.62
                      musicfee  $8,000.00     $7,426.45        $573.55
                      talent   $33,000.00    $31,411.37      $1,588.63
                               -----------   -----------    -----------
                               $101,000.00   $108,640.44    $-7,640.44
                               -----------   -----------    -----------

          Staff       fulltime $295,000.00   $293,988.43     $1,011.57
                      parttime $100,000.00    $99,869.08       $130.92
                               -----------   -----------    -----------
                               $395,000.00   $393,857.51     $1,142.49
                               -----------   -----------    -----------

          Travel      gas       $2,000.00     $1,522.19        $477.81
                      leases    $8,000.00     $6,934.80      $1,065.20
                               -----------   -----------    -----------
                               $10,000.00     $8,456.99      $1,543.01
                               -----------   -----------    -----------

                   4           ===========   ===========    ===========
                             4 $775,000.00   $780,011.28    $-5,011.28
                   4           ===========   ===========    ===========
```

The basic structure of this report differs from the structure of the reports you have seen so far in the following ways:

1. In the previous reports, each detail row represents one observation in the data set. In contrast, each detail row in this report consolidates all observations in the data set with a unique combination of values for 'Department' and 'Account'. In each case this data set contains two such observations, one for each quarter. The amounts displayed for the numeric variables in the report represent the sums of the values of those variables for both quarters. For instance, the first row of the report shows that the equipment department

   □ budgeted $80,000.00 for leases for the first and second quarters combined

   □ spent $80,000.00 for leases during the two quarters

   □ had a balance of zero for leases at the end of the two quarters.

2. In addition to consolidating data from multiple observations into one row of the report, this report summarizes data for each department. In this case the summaries appear after the last row for each department; however, you can easily place the summary before the first row for each department. A row of the report that summarizes other rows in the report is a *summary line*.

3. The summary line is one kind of *break line*. The dashed lines that separate the summary line from the body of the report are also break lines. As you will see shortly, you can customize break lines to suit your needs.

4. This report also includes three break lines at the end of the report. Two of these lines are, again, visual separators. The middle line in this group of break lines is a summary line for the entire report. This summary shows that altogether, the five departments in the company

   □ had a budget of $775,000.00 for the first and second quarters combined

   □ spent $780,011.28 during the two quarters

   □ overspent their budgets by a total of $5,011.28.

Forming a Column for Each Value of a Variable

The report in Output 1.6 rearranges the data and uses the same variable in multiple ways in one report.

Output 1.6
Using the Same Variable in Multiple Ways in One Report

```
                                                                     1
                                   ─────────QTR─────────
                               1    1st    1    2nd        3
         DEPT      ACCOUNT          BUDGET  2      BUDGET        BUDGET
         Equipment  lease      $40,000.00      $40,000.00   $80,000.00
                    maint      $10,000.00      $12,000.00   $22,000.00
                    purchase   $40,000.00      $20,000.00   $60,000.00
                    rental      $4,000.00       $6,000.00   $10,000.00
                    sets        $7,500.00       $7,500.00   $15,000.00
                    tape        $8,000.00      $12,000.00   $20,000.00
         Facilities rent       $24,000.00      $24,000.00   $48,000.00
                    supplies    $2,750.00       $2,750.00    $5,500.00
                    utils       $5,000.00       $3,500.00    $8,500.00
         Other      advert     $30,000.00      $30,000.00   $60,000.00
                    musicfee    $3,000.00       $5,000.00    $8,000.00
                    talent     $13,500.00      $19,500.00   $33,000.00
         Staff      fulltime  $130,000.00     $165,000.00  $295,000.00
                    parttime   $40,000.00      $60,000.00  $100,000.00
         Travel     gas           $800.00       $1,200.00    $2,000.00
                    leases      $3,500.00       $4,500.00    $8,000.00
```

Again, the basic structure of this report differs from that of previous reports as follows:

1. This report contains a column for each unique value of the variable QTR (formatted as 1st and 2nd).

2. The variable BUDGET appears below QTR, so that the values of BUDGET for each quarter appear side-by-side for each account within each department.

3. The report also uses the variable BUDGET independently of QTR, so that the last column of the report represents the sum of BUDGET for both quarters, just as it does in Output 1.5.

Consider, for example, the first row of the report. This row shows that the equipment department

□ budgeted $40,000.00 for leases in the first quarter

□ budgeted $40,000.00 for leases in the second quarter

□ budgeted a total of $80,000.00 for leases for both quarters.

Using Advanced Features

The report in Output 1.7 illustrates some advanced features of the REPORT procedure. Only two pages of the report appear in Output 1.7.

Output 1.7
Using Some
Advanced Features
of PROC REPORT

```
                                    ____Quarter____           3
                                  1    1st    1    2nd     Year-to-Date
          Department  Account       Balance  2   Balance      Balance
          Equipment   lease          $0.00         $0.00        $0.00
                      maint       $2,457.87     $1,324.71    $3,782.58
                      purchase   $-8,282.38     $2,230.85   $-6,051.53*  5
                      rental         $1.13       $517.06      $518.19
                      sets        $-842.68      $-579.62   $-1,422.30*
                      tape       $1,170.58       $573.27    $1,743.85

          ----------              -----------   -----------  -------------
          Equipment              $-5,495.48    $4,066.27   $-1,429.21*

          27% of the year-to-date budget is allocated to this department.  4
```

```
                                    ____Quarter____
                                       1st         2nd     Year-to-Date
          Department  Account       Balance     Balance      Balance
          Facilities  rent            $0.00       $0.00        $0.00
                      supplies      $533.45       $7.52      $540.97
                      utils         $776.71      $55.19      $831.90

          ----------              -----------  -----------  -------------
          Facilities             $1,310.16      $62.71    $1,372.87

          8% of the year-to-date budget is allocated to this department.
```

Note in particular the following features of this report:

1. Each unique value of 'Quarter' has a separate column in the report.

2. Values for 'Balance' appear below 'Quarter'; that is, all values of 'Balance' for the first quarter appear in one column, and all values of 'Balance' for the second quarter appear in another.

3. The report also contains a year-to-date balance for each account.

4. In addition to creating the kinds of break lines shown in Output 1.4, this report uses DATA step programming to customize break lines that provide information about each department.

5. The report contains an additional variable, a character variable, that flags accounts that are over budget with an asterisk(*). The space between this variable and the preceding column has been removed so that the asterisk appears immediately after the number it flags.

Customizing a Report

The report you create in the last tutorial in the book appears in Output 1.8. The report definition for this report extends the use of DATA step programming within PROC REPORT to further customize the report.

Output 1.8
Customizing a
Report

```
                                                                        1
          Quarter  Department        Balance       Budget       Actual
             1     Equipment      $-5,495.48  $109,500.00  $114,995.48
                   Facilities      $1,310.16   $31,750.00   $30,439.84
                   Other          $-1,514.21   $46,500.00   $48,014.21
                   Staff          $-1,492.80  $170,000.00  $171,492.80
                   Travel            $717.59    $4,300.00    $3,582.41
    1                                             2
The largest overdraw for this quarter was in the Equipment department.
It was overdrawn by $5,495.48.  3

             2     Equipment       $4,066.27   $97,500.00   $93,433.73
                   Facilities         $62.71   $30,250.00   $30,187.29
                   Other          $-6,126.23   $54,500.00   $60,626.23
                   Staff           $2,635.29  $225,000.00  $222,364.71
                   Travel            $825.42    $5,700.00    $4,874.58
                                                 2
The largest overdraw for this quarter was in the Other department.
It was overdrawn by $6,126.23.  3

                   ----------   ----------   ----------   ----------
                   Total:       $-5,011.28  $775,000.00  $780,011.28
                   ----------   ----------   ----------   ----------

============================================4 ============================5 ===========
The largest overdraw was in the Other department during the 2nd quarter.
It was overdrawn by $6,126.23.  6
=====================================================================================
```

In this report DATA step programming within PROC REPORT does the following:

1. customizes the text of the break lines that appear between quarters and at the end of the report

2. determines the department that most overspent its budget in each quarter

3. determines by how much this department was overdrawn

4. determines the department that was most overspent for the year to date

5. determines in which quarter this overspending occurred

6. determines by how much this department overspent its budget.

Chapter **2** Creating a Simple Report

Introduction

This chapter teaches you how to create a simple report with the REPORT procedure, how to store the definition of that report, and how to use the definition to create the same report for any SAS data set that contains the variables used in the report definition.

Understanding the SAS Data Set REPORT.BUDGET

The reports in this book are based on the SAS data set REPORT.BUDGET. This data set contains two quarters of financial data for a company that produces videotapes. The data set contains the following variables:

QTR   is the quarter of the year.

DEPT   is the name of a department in the company.

ACCOUNT   is the name of an account within a department.

BUDGET   is the amount of money budgeted for an account.

ACTUAL   is the amount of money spent by an account.

A simple listing of the data set, produced by the PRINT procedure, appears in Output 2.1. The DATA step that produces the data set is in the Appendix, "Creating the Data Set REPORT.BUDGET." The DATA step and the raw data are shipped with the SAS sample library. Contact your SAS Software Representative for information on the location of the sample library on your operating system.

Output 2.1
The SAS Data Set
REPORT.BUDGET

```
                              The SAS Data Set BUDGET                        1

        QTR    DEPT         ACCOUNT         BUDGET           ACTUAL

         1     Staff        fulltime     $130,000.00      $127,642.68
         2     Staff        fulltime     $165,000.00      $166,345.75
         1     Staff        parttime      $40,000.00       $43,850.12
         2     Staff        parttime      $60,000.00       $56,018.96
         1     Equipment    lease         $40,000.00       $40,000.00
         2     Equipment    lease         $40,000.00       $40,000.00
         1     Equipment    purchase      $40,000.00       $48,282.38
         2     Equipment    purchase      $20,000.00       $17,769.15
         1     Equipment    tape           $8,000.00        $6,829.42
         2     Equipment    tape          $12,000.00       $11,426.73
         1     Equipment    sets           $7,500.00        $8,342.68
         2     Equipment    sets           $7,500.00        $8,079.62
         1     Equipment    maint         $10,000.00        $7,542.13
         2     Equipment    maint         $12,000.00       $10,675.29
         1     Equipment    rental         $4,000.00        $3,998.87
         2     Equipment    rental         $6,000.00        $5,482.94
         1     Facilities   rent          $24,000.00       $24,000.00
         2     Facilities   rent          $24,000.00       $24,000.00
         1     Facilities   utils          $5,000.00        $4,223.29
         2     Facilities   utils          $3,500.00        $3,444.81
         1     Facilities   supplies       $2,750.00        $2,216.55
         2     Facilities   supplies       $2,750.00        $2,742.48
         1     Travel       leases         $3,500.00        $3,045.15
         2     Travel       leases         $4,500.00        $3,889.65
         1     Travel       gas              $800.00          $537.26
         2     Travel       gas            $1,200.00          $984.93
         1     Other        advert        $30,000.00       $32,476.98
         2     Other        advert        $30,000.00       $37,325.64
         1     Other        talent        $13,500.00       $12,986.73
         2     Other        talent        $19,500.00       $18,424.64
         1     Other        musicfee       $3,000.00        $2,550.50
         2     Other        musicfee       $5,000.00        $4,875.95
```

Some reports use the entire data set; others use only some of the data. In the latter cases the data set option OBS= or the WHERE statement subsets the data.

Creating a Report with the PROMPT Facility

The REPORT procedure provides a PROMPT facility, implemented through the PROMPTER window, to help you become familiar with the process of building a report. The PROMPT facility is a tool that prompts you for information as you add either data set variables or statistics to a report. The PROMPT facility steps you through the most commonly used parts of the windows you would use to add either a data set variable or a statistic to the report. However, the PROMPTER window provides more explanation than the other windows provide. You choose whether or not to use this tool.

This chapter uses the PROMPT facility to create a report. After you complete this report, you should know how to

□ invoke the REPORT procedure with the PROMPT facility

□ make selections in the report and in windows

□ add data set variables to the report

□ use PROC REPORT's help facility to get help with a choice in a window

□ store a report definition

□ terminate the REPORT procedure

□ create the same report for any data set that contains the variables in the report definition.

Building Your First Report

The report you produce in this tutorial is the simple report in Output 2.2. This report shows the amounts budgeted for and actually spent by each account within each department during each quarter. Producing this simple report teaches you how to use some of the basic features of the REPORT procedure. In later tutorials you learn how to make more sophisticated reports.

Output 2.2
Your First Report

```
                                  The SAS System                        1

                                      Amount        Amount
         Quarter  Department  Account  Budgeted       Spent
               1  Equipment   lease     $40,000.00   $40,000.00
                              maint     $10,000.00    $7,542.13
                              purchase  $40,000.00   $48,282.38
                              rental     $4,000.00    $3,998.87
                              sets       $7,500.00    $8,342.68
                              tape       $8,000.00    $6,829.42
                  Facilities  rent      $24,000.00   $24,000.00
                              supplies   $2,750.00    $2,216.55
                              utils      $5,000.00    $4,223.29
                  Other       advert    $30,000.00   $32,476.98
                              musicfee   $3,000.00    $2,550.50
                              talent    $13,500.00   $12,986.73
                  Staff       fulltime $130,000.00  $127,642.68
                              parttime  $40,000.00   $43,850.12
                  Travel      gas          $800.00      $537.26
                              leases     $3,500.00    $3,045.15
               2  Equipment   lease     $40,000.00   $40,000.00
                              maint     $12,000.00   $10,675.29
                              purchase  $20,000.00   $17,769.15
                              rental     $6,000.00    $5,482.94
                              sets       $7,500.00    $8,079.62
                              tape      $12,000.00   $11,426.73
                  Facilities  rent      $24,000.00   $24,000.00
                              supplies   $2,750.00    $2,742.48
                              utils      $3,500.00    $3,444.81
                  Other       advert    $30,000.00   $37,325.64
                              musicfee   $5,000.00    $4,875.95
                              talent    $19,500.00   $18,424.64
                  Staff       fulltime $165,000.00  $166,345.75
                              parttime  $60,000.00   $56,018.96
                  Travel      gas        $1,200.00      $984.93
                              leases     $4,500.00    $3,889.65
```

Notice that the report orders the rows first by their values of 'Quarter', then by their values of 'Department' within 'Quarter', and finally by their values of 'Account' within 'Department'.

Invoking the REPORT Procedure with the PROMPT Facility

CMS, MVS, VSE ....

If you are using the SAS Display Manager System on the CMS, MVS, or VSE operating system, issue the following command from the command line of the PROGRAM EDITOR window before invoking the REPORT procedure:

```
nulls off
```

Issue this command each time you start a SAS session in which you intend to use the REPORT procedure. Instructions for turning NULLS off when you aren't using display manager appear later in this chapter.

....

To invoke the REPORT procedure with the PROMPT facility, submit the following SAS statements:

```
libname report 'SAS-data-library';

proc report data=report.budget prompt;
run;
```

where *SAS-data-library* is your host system's name for the SAS data library that contains the data set BUDGET. The PROMPT option opens the REPORT window and starts the PROMPT facility.

After you submit these statements, four windows appear on your display, as Display 2.1 illustrates.

Display 2.1
Initial Display of
Windows with
PROMPT Facility

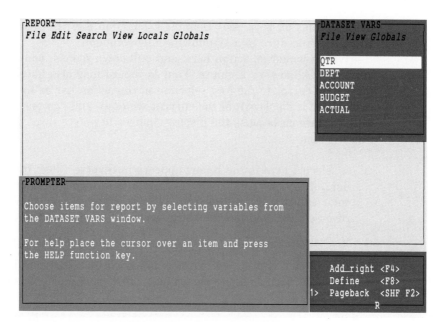

The four windows are described below:

REPORT window
 contains the text of the report. In this example the REPORT window is
 initially empty.

DATASET VARS window
 lists the variables in the data set.

PROMPTER window
 prompts you on how to proceed.

RKEYS window
 displays the default definitions for 12 function keys defined as REPORT
 commands. (These definitions are system dependent.) In Display 2.1, the
 PROMPTER window blocks most of the RKEYS window, including the
 window name.

Making Selections in the REPORT Procedure

As you create a report, you select items from the report, from windows, from
action bars across the tops of windows, and from pull-down menus that appear
when you select from action bars. To make a selection, move the cursor to the
desired location and press ENTER, RETURN, or the appropriate mouse button.

 Within the report itself, use the cursor keys or a mouse to move the cursor. If
you aren't using a mouse, you may find the NEXTFIELD and PREVFIELD
commands useful for moving from place to place. On some operating systems,
these two commands are assigned to function keys by default. Use the KEYS
command (under **Globals** in the action bar at the top of the DATASET VARS
window) to determine the commands associated with your function keys.

 Note: The KEYS command is available under **Globals** in the action bar of
any window with an action bar, but at this point in the tutorial, if you want to
issue the KEYS command, you must use the action bar in the DATASET VARS
window because the REPORT window is not currently active. For information on

defining function keys, see "Defining a Function Key" in Chapter 7, "Using the REPORT Procedure in a Windowing Environment."

Within windows, action bars, and pull-down menus, you can use the cursor keys, the TAB key, or a mouse. Your keyboard may also have a HOME key that moves the cursor to the first selection in the action bar or to the command line (whichever is displayed) of the current window. You can tell when you have selected an item because the display changes in one of three ways to highlight the selected item:

□ A selected item in the report appears in reverse video; that is, for the selected item, the background color appears in the foreground and the foreground color appears in the background. Selected items in certain windows appear in reverse video and move toward the top of the list of items.

□ If a selected item in a window is preceded by a circle, that circle is filled.

□ If a selected item in a window is preceded by a box, that box contains a character such as a checkmark (√), an X, or a greater-than sign (>).

Adding Data Set Variables to the Report

CMS, MVS, and VSE ....

If you are using the CMS, MVS, or VSE operating system but you did not invoke PROC REPORT from display manager, you need to turn NULLS off before proceeding. To do so, follow the directions below:

1. Move the cursor to the action bar at the top of the DATASET VARS window and select **Globals**.

2. From the pull-down menu that appears, select **Command**.

3. On the command line, type `nulls off` and press ENTER.

4. On the command line, type `command` and press ENTER.

....

Now select **QTR**, **DEPT**, **ACCOUNT**, **BUDGET**, and **ACTUAL** from the DATASET VARS window. The cursor is already on **QTR**, so just press ENTER (or RETURN or the appropriate mouse button). **QTR** is now highlighted. If necessary, move the cursor to the next variable in the list, **DEPT**, and press ENTER. Repeat this process until you have selected all five variables.

Note: In some windowing environments the cursor automatically moves to the next item in the list when you make a selection from the DATASET VARS window. In other windowing environments, it doesn't.

After you select all the variables, your display looks like Display 2.2.

Note: The order in which you select the variables determines the order in which they appear in the report. PROC REPORT builds the report from left to right, starting with the first variable in the list and proceeding to the bottom of the list. Be sure to select the variables in the order shown.

Display 2.2
Selecting Variables
from the
DATASET VARS
Window

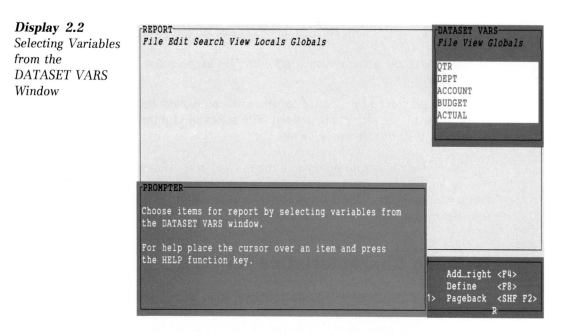

Now that you have selected the variables to use in the report, you are ready
to exit the DATASET VARS window. Move the cursor to **File** in the action bar at
the top of the DATASET VARS window, and press ENTER. Select **End** from the
pull-down menu that appears, as shown in Display 2.3.

Display 2.3
Exiting a Window

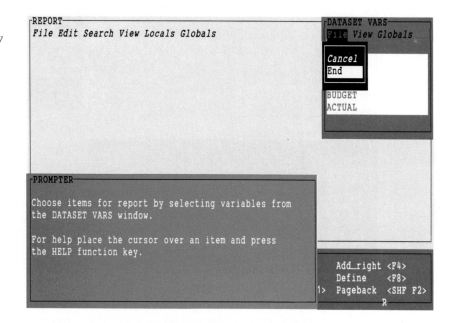

Note: You always have at least two commands to choose from when you exit
a window: the OK command and the CANCEL command. The OK command (also
called the END command) implements the choices and the changes in values you
made while in that window. Typically, the CANCEL command cancels any changes
you have made in the window since you opened it. In the PROMPTER window,
however, the CANCEL command cancels prompting at that point.

The OK and CANCEL commands are in one of two places:

□ in the pull-down menu associated with **File** in the action bar at the top of the window.

□ in push buttons at the bottom of the window. A *push button* is a part of a window that is always highlighted. The selection of a push button initiates an action, such as exiting a window.

Particularly if you do not have a mouse, you may want to define a function key as the END command and another as the CANCEL command so that you can exit a window with a single keystroke. For information on defining function keys, see "Defining a Function Key" in Chapter 7.

Defining Characteristics for QTR

The display of the PROMPTER window now changes to a prompt for how you want to use the first variable you selected (QTR) in the report. As you can see in Display 2.4, the name displayed in the PROMPTER window changes. The current name, DEFINITION, is the name of the window that you would use to supply this information if you were not using the PROMPT facility.

Display 2.4
Prompting for
Usage of QTR

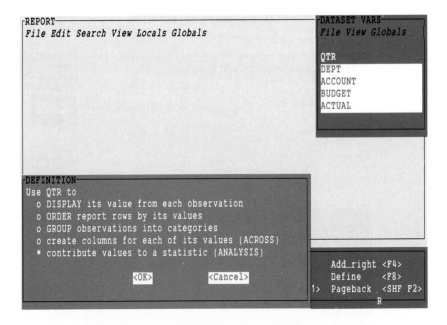

In this report, you are going to show all the information for the first quarter before the information for the second quarter. Thus, you want to order the rows of the report according to values of QTR. Select **ORDER report rows by its values** from the window.

Note: To make a selection, position the cursor anywhere in the field containing the text. If an empty circle precedes the text, you can also place the cursor on the circle.

After making this selection, select the **OK** push button.

Next, as Display 2.5 illustrates, the window prompts you for the format, width, and column header for QTR. Initially, the default values appear in the window.

Display 2.5
Prompting for
Format, Width,
and Column
Header for QTR

```
┌REPORT─────────────────────────────────────────┐  ┌DATASET VARS────────┐
│ File Edit Search View Locals Globals           │  │ File View Globals   │
│                                                │  │                     │
│                                                │  │ QTR                 │
│                                                │  │ DEPT                │
│                                                │  │ ACCOUNT             │
│                                                │  │ BUDGET              │
│                                                │  │ ACTUAL              │
│                                                │  │                     │
│                                                │  │                     │
│                                                │  │                     │
│ ┌DEFINITION──────────────────────────────────┐ │  │                     │
│ │ Supply new attributes for QTR by typing over│ │  │                     │
│ │ existing values below.                      │ │  │                     │
│ │                                             │ │  │                     │
│ │    Format  =   BEST9.                       │ │  └─────────────────────┤
│ │    Width   =   9                            │ │  │                     │
│ │    Header  =   QTR                          │ │  │  Add_right  <F4>    │
│ │                                             │ │  │  Define     <F8>    │
│ │      <OK>       <Backup>      <Cancel>      │ │  │1>Pageback  <SHF F2> │
│ └─────────────────────────────────────────────┘ │  │            R        │
└──────────────────────────────────────────────────────────────────────────┘
```

To enter the values you want, move the cursor to the appropriate field and type the value you want to use. It is important to note that these values apply only to the report. They do not affect the data set. Make the following changes to the attributes for QTR:

□ Move the cursor to the **Format** field and type the **1.** format. Be sure to include the period as part of the format.

□ Move the cursor to the **Width** field and type **7** to make the field just wide enough to accommodate the column header.

□ Type **Quarter** in the **Header** field.

When your display looks like Display 2.6, select the **OK** push button.

Display 2.6
Setting
Characteristics for
QTR

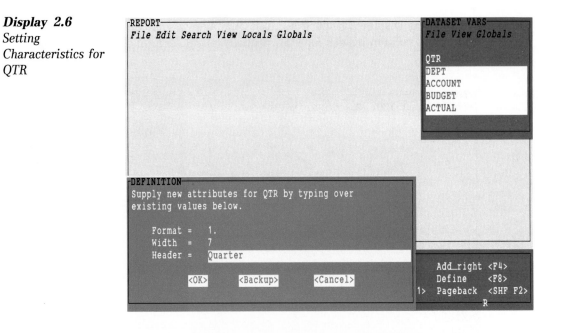

Note: In some of the windows of the PROMPT facility, you have three commands to choose from when you exit a window. In addition to the OK and CANCEL commands, you can select the BACKUP command. If you do so, the REPORT procedure backs up to the previous PROMPTER window.

The next prompt, which appears in Display 2.7, asks if you want to calculate summary statistics or subtotals for the variable QTR.

Display 2.7
Prompting for
Statistics and
Subtotals

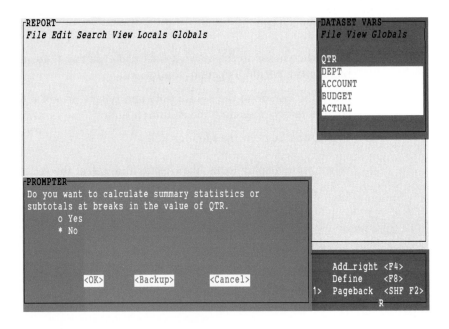

This first report uses neither of these features, although you will learn how to use both of them in Chapter 4, "Grouping and Summarizing Data." The default selection in this window is **No,** and the cursor is on the **OK** push button, so simply press ENTER.

Defining Characteristics for DEPT

The REPORT procedure now prompts you for the same set of characteristics for the next variable, DEPT. First, as you can see in Display 2.8, the procedure offers choices for how to use the variable in the report.

Display 2.8
Prompting for
Usage for DEPT

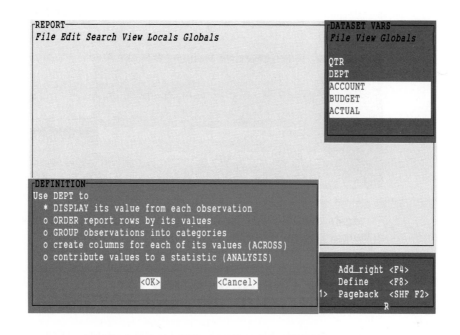

A look at Output 2.2 shows that within the sections of the report for each quarter, PROC REPORT has ordered the rows by their values of DEPT. So, once again, select **ORDER report rows by its value**. Then select the **OK** push button.

In the next window, leave the value of **Format** at **$10.**, the default. Leave the value of **Width** at **10**. Type in **Department** for the column header.

When your display looks like Display 2.9, select the **OK** push button.

Display 2.9
*Prompting for
Format, Width,
and Column
Header for DEPT*

```
┌REPORT──────────────────────────────────────────────────┐  ┌DATASET VARS─────┐
│ File Edit Search View Locals Globals                    │  │ File View Globals│
│                                                         │  │                  │
│                                                         │  │ QTR              │
│                                                         │  │ DEPT             │
│                                                         │  │ ACCOUNT          │
│                                                         │  │ BUDGET           │
│                                                         │  │ ACTUAL           │
│                                                         │  │                  │
│                                                         │  │                  │
│ ┌DEFINITION───────────────────────────────────────┐    │  │                  │
│ │ Supply new attributes for DEPT by typing over    │    │  │                  │
│ │ existing values below.                           │    │  │                  │
│ │                                                  │    │  │                  │
│ │     Format =   $10.                              │    │  │                  │
│ │     Width  =   10                                │    │  ┌──────────────────┐
│ │     Header =  │Department                   │    │    │  │ Add_right <F4>   │
│ │                                                  │    │  │ Define    <F8>   │
│ │        <OK>      <Backup>      <Cancel>          │    │ 1>│ Pageback <SHF F2>│
│ └──────────────────────────────────────────────────┘    │  │                R │
└─────────────────────────────────────────────────────────┘  └──────────────────┘
```

Again, the PROMPT facility asks about statistics and subtotals. Accept the default of **No** and select the **OK** push button.

Defining Characteristics for ACCOUNT, BUDGET, and ACTUAL

Continue through this same process for the variables ACCOUNT, BUDGET, and ACTUAL. Make the following selections for ACCOUNT:

□ Use the ACCOUNT variable to **ORDER report rows by its values.**

□ In the next window,

  □ accept the default values for **Format** and **Width**

  □ type in `Account` for the column header.

□ Once again, select **No** from the final prompt.

Make the following selections for BUDGET:

□ Use the variable BUDGET to **DISPLAY its value from each observation.**

□ Accept the default values for **Format** and **Width.** (Note that the default format and width result from the format of DOLLAR11.2 associated with BUDGET in the data set.)

□ Type in `Amount/Budgeted` for the column header. Here, the slash character (/) is the default *split character*. If you use the split character in a column header, the REPORT procedure breaks the header when it reaches that character and continues the header on the next line. The split character itself is not part of the column header. Thus, the column header you specify here occupies two lines in the report. The first line contains the characters that precede the split character. The second line contains those that follow the split character. (For information on changing the split character, see the documentation for the ROPTIONS window and for the SPLIT= option in the PROC REPORT statement.)

Note: By choosing to use BUDGET to display its value from each observation, you rule out the options of calculating statistics and subtotals.

Make the same choices for ACTUAL that you made for BUDGET, except type in `Amount/Spent` for the column header.

Exiting the PROMPTER Window

You have now specified the characteristics for all the variables you selected from the DATASET VARS window. As Display 2.10 shows, the PROMPT facility asks if you want to add more items to the report.

Display 2.10
Prompting for
More Items to Add
to the Report

```
┌REPORT─────────────────────────────────────────────────────────────┐
│ File Edit Search View Locals Globals                               │
│                                                                    │
│                          The SAS System                          1 │
│                                                                    │
│                                     Amount       Amount            │
│            Quarter  Department  Account  Budgeted      Spent        │
│                  1  Equipment   lease   $40,000.00  $40,000.00      │
│                                 maint   $10,000.00   $7,542.13      │
│                                 purchase $40,000.00  $48,282.38     │
│                                 rental   $4,000.00   $3,998.87      │
│                                 sets     $7,500.00   $8,342.68      │
│                                 tape     $8,000.00   $6,829.42      │
│┌PROMPTER──────────────────────────────────────────┐$24,000.00      │
││You can now modify the report directly or         │ $2,216.55      │
││continue to add items with prompting.             │ $4,223.29      │
││Do you want to add more items?                    │$32,476.98      │
││                                                  │ $2,550.50      │
││    o Yes                                         │                │
││    * No                                          ├────────────────┤
││                                                  │ Add_right <F4> │
││        <OK>            <Cancel>                  │ Define    <F8> │
││                                                  │1> Pageback <SHF F2>│
│└──────────────────────────────────────────────────┘        R       │
└────────────────────────────────────────────────────────────────────┘
```

At this point, if you had not originally selected all the variables you wanted from the DATASET VARS window, you could continue to add variables by selecting **Yes**. However, you have already added all the variables to the report, so accept the default of **No** and select the **OK** push button.

The PROMPTER window disappears, and the REPORT procedure displays the report in Display 2.11.

Display 2.11
Report with All
the Data Set
Variables

```
┌REPORT──────────────────────────────────────────────────────────────┐
│ File Edit Search View Locals Globals                                │
│                                                                     │
│                         The SAS System                            1 │
│                                                                     │
│                                         Amount        Amount        │
│              Quarter  Department  Account    Budgeted        Spent  │
│                    1  Equipment   lease    $40,000.00   $40,000.00   │
│                                   maint    $10,000.00    $7,542.13   │
│                                   purchase $40,000.00   $48,282.38   │
│                                   rental    $4,000.00    $3,998.87   │
│                                   sets      $7,500.00    $8,342.68   │
│                                   tape      $8,000.00    $6,829.42   │
│                       Facilities  rent     $24,000.00   $24,000.00   │
│                                   supplies  $2,750.00    $2,216.55   │
│                                   utils     $5,000.00    $4,223.29   │
│                       Other       advert   $30,000.00   $32,476.98   │
│                                   musicfee  $3,000.00    $2,550.50   │
│                                                                     │
├RKEYS────────────────────────────────────────────────────────────────┤
│ Add_above <F1>    Add_below <F2>    Add_left  <F3>    Add_right <F4>  │
│ Break     <F5>    CGrow     <F6>    CShrink   <F7>    Define    <F8>  │
│ Delete    <F9>    Move    <SHF F0>  Page    <SHF F1>  Pageback <SHF F2>│
│                                                                   R  │
└─────────────────────────────────────────────────────────────────────┘
```

Scrolling and Paging through the Report

The report in Display 2.11 is the same as the report in Output 2.2, but the amount of the report you can see depends on the size of your display and the line size and page size you are using in your SAS session.

Note: You can alter the page size and the line size of the report with the **Pagesize** and the **Linesize** attributes in the ROPTIONS window.

Use the commands under **View** in the action bar at the top of the REPORT window to view different parts of the report. The PAGE command displays the next page of the report. The PAGEBACK command displays the previous page. The FORWARD, BACKWARD, RIGHT, and LEFT commands display text that is on the same page of the report but is not visible on your display.

By default the REPORT procedure defines function keys for the PAGE and PAGEBACK commands. Particularly if you don't have a mouse, you may want to define function keys for the other commands as well.

When you have viewed the report to your satisfaction, use the PAGEBACK, BACKWARD, and LEFT commands as necessary to return to the top of the report.

Storing a Report Definition

As you work, the REPORT procedure maintains a definition of the current report. You can store that definition in a catalog entry of type REPT and use it to create a similar report for any data set that contains variables with the same names as the ones used in the report definition.

To store the current report definition, first select **File** from the action bar at the top of the REPORT window. Display 2.12 shows the pull-down menu that appears.

Display 2.12
Pull-Down Menu for File in the REPORT window

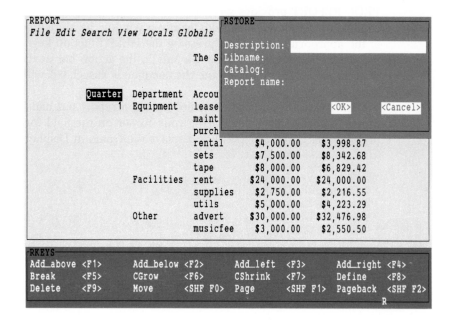

Note: If you make the wrong selection from an action bar, simply move the cursor out of the pull-down menu that appears and press ENTER, RETURN, or the appropriate mouse button.

From this pull-down menu, select **RSTORE**. The RSTORE command displays the RSTORE window, which appears in Display 2.13. In the RSTORE window, you specify where you want to store the report definition.

Display 2.13
RSTORE Window

The cursor is at the **Description** field. The text you enter here is the description that appears with the catalog entry when you use the RLOAD window to load a report definition or when you use the CATALOG procedure or the CATALOG window to see the contents of the catalog containing this report definition. A description can contain up to 40 characters. Type `tutorials: chapter 2` in the **Description** field.

In the **Libname** field you specify a libref pointing to the SAS data library that contains the catalog in which you want to store the report. The SAS System always defines one libref, SASUSER, for you, so type `sasuser` in this field.

In the **Catalog** field, you specify the name of the catalog in which to store the report. If the catalog does not already exist, the SAS System creates it for you. Type `tutors` in the **Catalog** field.

The **Report name** field specifies the name of the entry that will contain this report definition. Enter `chap2` in this field.

Note: You do not need to specify the entry type, REPT, because all report definitions have the same entry type.

When you have filled in all the fields, select the **OK** push button. The REPORT procedure stores the report in the catalog entry SASUSER.TUTORS.CHAP2.REPT and sends you a message telling you it has done so. If the catalog is new, PROC REPORT issues a warning in the message window telling you that you have created a new catalog. To exit this window, select **File** from the action bar in the MESSAGE window. Select either **Cancel** or **End** from the pull-down menu that appears.

Note: As you develop a report, it is wise to save report definitions as you progress toward your final report. Thus, if you make a mistake or if you have problems with your host system, a recent version of the report is available. Saving report definitions can save you a lot of work if, for instance, you experience a system failure while you are developing a report.

Using the Help Facility

PROC REPORT provides an extensive online help facility. You can get help with commands, windows, and fields within a window by simply positioning the cursor at the appropriate place and pressing the HELP function key. (Remember, you can use the KEYS command under **Globals** in the action bar at the top of any window with an action bar to determine the commands associated with your function keys.)

For example, suppose you were storing a report but had forgotten what to put in the **Report name** field. Placing your cursor on the field for **Report name** and pressing the HELP function key displays the frame in Display 2.14.

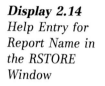

Display 2.14
Help Entry for
Report Name in
the RSTORE
Window

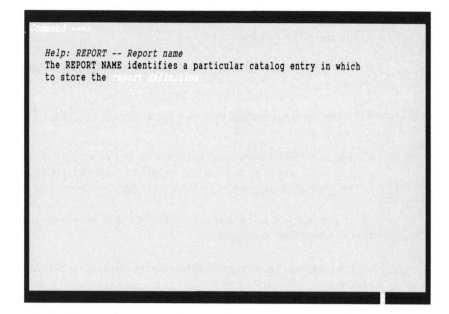

Note: Some help frames include words or groups of words, like **report definition** in this frame, that are highlighted in some way. The exact method of highlighting varies from one operating system to another. Typically, the text is in a different color from most of the text, or in reverse video. You can get help on these topics by moving the cursor to the highlighted text you are interested in and pressing the HELP function key, ENTER, RETURN, or the corresponding mouse button.

To return to your report, type **end** at the command line of the HELP frame, and press ENTER.

Terminating the REPORT Procedure

Now that you've created and stored your first report definition, you're ready to terminate the REPORT procedure and print a hard copy of the report.

To terminate the procedure, select **File** from the action bar at the top of the report window. When the pull-down menu appears, select **Quit.** PROC REPORT opens a dialog box that asks you to confirm or cancel the QUIT command. Select the **OK** push button to confirm the command.

Printing the Report

If you are using the SAS Display Manager System, follow these steps to print your report:

1. Submit the following SAS statements from the PROGRAM EDITOR window to produce the report:

   ```
   proc report data=report.budget report=sasuser.tutors.chap2 windows;
   run;
   ```

   After the SAS System executes these statements, it displays the REPORT window.

2. Use the PRINT command (under **File** in the action bar at the top of the REPORT window) to send the report to your default printer. For information on your default printer, contact your SAS Software Representative.

   If you are not using display manager, route your output to a file and submit the following SAS statements:

   ```
   proc report data=report.budget report=sasuser.tutors.chap2 nowindows;
   run;
   ```

   For information on routing output to a file, see the SAS documentation for your host system.

Using the REPORT Language to Produce the Report

The following SAS statements produce the report you created in this tutorial. For a detailed explanation of these statements, see "Example 1: Ordering the Rows of a Report" in Chapter 10, "The REPORT Language."

```
proc report data=report.budget;

   column qtr dept account budget actual;

   define qtr     / order format=1. width=7 'Quarter';
   define dept    / order format=$10. width=10 'Department';
   define account / order format=$8. width=8 'Account';
   define budget  / display format=dollar11.2 width=11
                            'Amount/Budgeted';
   define actual  / display format=dollar11.2 width=11
                            'Amount/Spent';
run;
```

Chapter 3 Enhancing Your First Report

Introduction

Now that you know how to produce a simple report, you are ready to try a slightly more complicated one. In this chapter you will create the report in Output 3.1.

Output 3.1
Enhanced Report

```
                            Report of Departments                    1
                                 by Quarter

                               Amount        Amount
      Quarter Department Account   Budgeted        Spent   Balance
      ------------------------------------------------------------

          1 Equipment  lease    $40,000.00   $40,000.00      $0.00
                       maint    $10,000.00    $7,542.13  $2,457.87
                       purchase $40,000.00   $48,282.38 $-8,282.38
                       rental    $4,000.00    $3,998.87      $1.13
                       sets      $7,500.00    $8,342.68   $-842.68
                       tape      $8,000.00    $6,829.42  $1,170.58
            Facilities rent     $24,000.00   $24,000.00      $0.00
                       supplies  $2,750.00    $2,216.55    $533.45
                       utils     $5,000.00    $4,223.29    $776.71
            Other      advert   $30,000.00   $32,476.98 $-2,476.98
                       musicfee  $3,000.00    $2,550.50    $449.50
                       talent   $13,500.00   $12,986.73    $513.27
            Staff      fulltime $130,000.00  $127,642.68  $2,357.32
                       parttime $40,000.00   $43,850.12 $-3,850.12
            Travel     gas         $800.00      $537.26    $262.74
                       leases    $3,500.00    $3,045.15    $454.85
```

```
                        Report of Departments                      2
                             by Quarter

                                  Amount      Amount
      Quarter  Department  Account  Budgeted      Spent    Balance
      -----------------------------------------------------------------

         2  Equipment   lease     $40,000.00  $40,000.00      $0.00
                        maint     $12,000.00  $10,675.29  $1,324.71
                        purchase  $20,000.00  $17,769.15  $2,230.85
                        rental     $6,000.00   $5,482.94    $517.06
                        sets       $7,500.00   $8,079.62   $-579.62
                        tape      $12,000.00  $11,426.73    $573.27
            Facilities  rent      $24,000.00  $24,000.00      $0.00
                        supplies   $2,750.00   $2,742.48      $7.52
                        utils      $3,500.00   $3,444.81     $55.19
            Other       advert    $30,000.00  $37,325.64  $-7,325.64
                        musicfee   $5,000.00   $4,875.95    $124.05
                        talent    $19,500.00  $18,424.64  $1,075.36
            Staff       fulltime $165,000.00 $166,345.75  $-1,345.75
                        parttime  $60,000.00  $56,018.96  $3,981.04
            Travel      gas        $1,200.00     $984.93    $215.07
                        leases     $4,500.00   $3,889.65    $610.35
```

This report includes several enhancements over the first report:

□ The report contains a variable, BALANCE, that is not in the data set.

□ The rows for each quarter appear on separate pages.

□ A blank line and underlining separate the column headers from the text of the report.

□ The title describes the contents of the report.

Invoking the REPORT Procedure with a Report Definition

After you completed your first report, you stored the report definition in the catalog entry SASUSER.TUTORS.CHAP2.REPT. Because the second report is an extension of the first, you can save time by starting your PROC REPORT session with that report definition. To do so, submit the following SAS statements:

```
proc report data=report.budget report=sasuser.tutors.chap2
          windows;
run;
```

As you saw in Chapter 2, "Creating a Simple Report," the DATA= option specifies the SAS data set to use in the report. The REPORT= option specifies the report definition to use. You don't need to specify the entry type because all report definitions are type REPT. The WINDOWS option opens the REPORT window without the PROMPT facility. Your display now looks like Display 3.1.

Display 3.1
Invoking the
REPORT
Procedure with a
Stored Report
Definition

```
┌REPORT──────────────────────────────────────────────────────────────────┐
│File Edit Search View Locals Globals                                     │
│                                                                          │
│                            The SAS System                            1   │
│                                                                          │
│                                         Amount        Amount             │
│                Quarter  Department  Account    Budgeted       Spent      │
│                   1     Equipment   lease    $40,000.00   $40,000.00     │
│                                     maint    $10,000.00    $7,542.13     │
│                                     purchase $40,000.00   $48,282.38     │
│                                     rental    $4,000.00    $3,998.87     │
│                                     sets      $7,500.00    $8,342.68     │
│                                     tape      $8,000.00    $6,829.42     │
│                         Facilities  rent     $24,000.00   $24,000.00     │
│                                     supplies  $2,750.00    $2,216.55     │
│                                     utils     $5,000.00    $4,223.29     │
│                         Other       advert   $30,000.00   $32,476.98     │
│                                     musicfee  $3,000.00    $2,550.50     │
│                                                                          │
└──────────────────────────────────────────────────────────────────────────┘
┌RKEYS─────────────────────────────────────────────────────────────────────┐
│Add_above <F1>      Add_below <F2>      Add_left  <F3>      Add_right <F4>  │
│Break     <F5>      CGrow     <F6>      CShrink   <F7>      Define    <F8>  │
│Delete    <F9>      Move    <SHF F0>    Page    <SHF F1>    Pageback <SHF F2>│
│                                                                        R   │
└───────────────────────────────────────────────────────────────────────────┘
```

Adding a Computed Variable

You can use items in a report to add computed variables to a report. The variables you compute become part of the report, but they are not part of the data set. In this section you compute a variable, BALANCE, that is the difference between the amount budgeted for an account and the amount spent.

First, select the item next to which you want to add the computed variable. In this report, you want to add the new variable to the right of ACTUAL, which is in the column labeled **Amount Spent**. Select this column by moving the cursor to the column header and pressing ENTER, RETURN, or the appropriate mouse button.

Next, issue the command that adds the variable to the report, in this case the ADD_RIGHT command.

To issue the ADD_RIGHT command, do one of the following:

□ Consult the RKEYS window and press the ADD_RIGHT function key.

□ Select **Edit** from the action bar at the top of the REPORT window. From the pull-down menu that appears, select **ADD_RIGHT**.

□ In the RKEYS window, select **ADD_RIGHT** or the name of the corresponding function key (for example, **F4**).

The ADDING window appears on your display, as Display 3.2 illustrates. Notice the header in this window that tells you that you are adding to the right of ACTUAL.

Display 3.2
ADDING Window

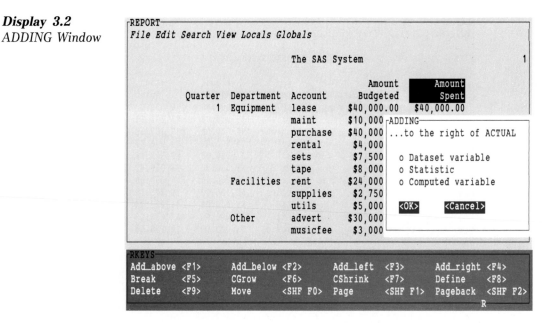

The ADDING window presents three choices: **Dataset variable, Statistic,** and **Computed variable.** Select **Computed variable.** When you have done so, select the **OK** push button.

The REPORT procedure now displays the COMPUTED VAR window, which appears in Display 3.3.

Display 3.3
COMPUTED VAR Window

Type `balance` at the prompt for the variable's name. Because this variable is numeric, you can ignore the **Character data** check box and the prompt for a length. You will use those features in Chapter 5, "Using Some Advanced Features of the Report Procedure." Next, select the **Edit Program** push button. PROC REPORT displays the COMPUTE window, which appears in Display 3.4.

Note: If you inadvertently select the **OK** push button instead of the **Edit Program** push button, you return to the REPORT window. Your report contains a column header for BALANCE, but no values. To get back to where you should be, select **BALANCE**, issue the DEFINE command (under **Edit** in the action bar at the top of the REPORT window), and from the DEFINITION window, select the **Edit Program** push button.

Display 3.4
COMPUTE
Window

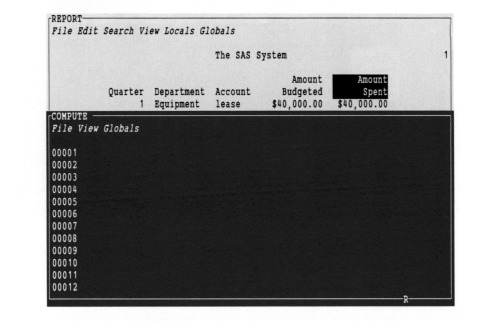

Once in the COMPUTE window, you define the variable BALANCE with DATA step statements. Enter the following assignment statement on the first line of the COMPUTE window:

```
balance=budget-actual;
```

Note: In the COMPUTE window, you refer to variables by their names, not by their column headers.

When you are ready, select **File** from the action bar at the top of the COMPUTE window. Select **End** from the pull-down menu that appears, as Display 3.5 illustrates.

Display 3.5
Exiting the
COMPUTE
Window

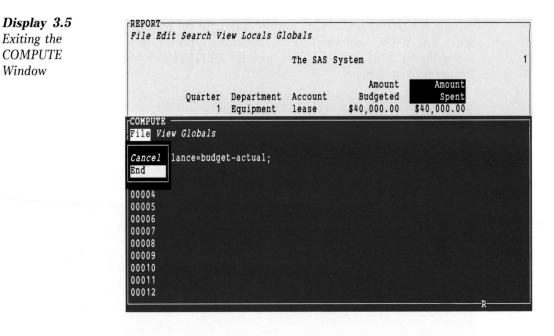

Now you are back at the COMPUTED VAR window. Select the **OK** push button to complete the process of adding BALANCE to the report. The new version of the report appears in Display 3.6.

Display 3.6
Report with the
Computed Variable
BALANCE

```
┌REPORT──────────────────────────────────────────────────────────────────┐
│ File Edit Search View Locals Globals                                    │
│                                                                         │
│                            The SAS System                           1   │
│                                              ┌────────┐                  │
│                                     Amount   │ Amount │                  │
│         Quarter  Department  Account  Budgeted│  Spent │  BALANCE        │
│            1     Equipment  lease   $40,000.00 $40,000.00        0       │
│                             maint   $10,000.00  $7,542.13     2458       │
│                             purchase $40,000.00 $48,282.38    -8282      │
│                             rental   $4,000.00  $3,998.87        1       │
│                             sets     $7,500.00  $8,342.68     -843       │
│                             tape     $8,000.00  $6,829.42     1171       │
│                  Facilities rent    $24,000.00 $24,000.00        0       │
│                             supplies  $2,750.00  $2,216.55     533       │
│                             utils     $5,000.00  $4,223.29     777       │
│                  Other      advert  $30,000.00 $32,476.98    -2477       │
│                             musicfee  $3,000.00  $2,550.50     450       │
│                                                                         │
├RKEYS────────────────────────────────────────────────────────────────────┤
│ Add_above <F1>      Add_below <F2>      Add_left  <F3>    Add_right <F4>  │
│ Break     <F5>      CGrow     <F6>      CShrink   <F7>    Define    <F8>  │
│ Delete    <F9>      Move    <SHF F0>    Page    <SHF F1>  Pageback <SHF F2>│
│                                                                     R    │
└─────────────────────────────────────────────────────────────────────────┘
```

Defining the Characteristics of a Variable

When you add a numeric computed variable to a report, the REPORT procedure uses the default column width (9 unless you change it in the ROPTIONS window) and the BESTw. format that spans the column. For instance, in this case the definition of BALANCE includes a width of 9 and a format of BEST9. In this report, however, you want the format and column width for BALANCE to match those for the variables BUDGET and ACTUAL.

Each item in the report is defined in a specific way. You can see the definition for an item by selecting that item and issuing the DEFINE command. So, to set the width and format for BALANCE, select **BALANCE** in the REPORT window and issue the DEFINE command in one of the following ways:

□ Consult the RKEYS window and press the DEFINE function key.

□ Select **Edit** from the action bar at the top of the REPORT window. From the pull-down menu that appears, select **DEFINE**.

□ In the RKEYS window, select **DEFINE** or the name of the corresponding function key (for example, **F8**).

The REPORT procedure displays the DEFINITION window, which appears in Display 3.7. The **Format** and **Width** fields appear in the column labeled **Attributes**.

Display 3.7
DEFINITION
Window for
BALANCE

```
┌REPORT─────────────────────────────────────────────────────────────────┐
│ File Edit Search View Locals Globals                                    │
│                                                                         │
│                              The SAS System                          1  │
│                                                                         │
│         Quarter┌DEFINITION──────────────────────────────────────────┐   │
│              1 │              Definition of BALANCE                  │   │
│                │   Usage      Attributes              Options  Color │   │
│                │   o DISPLAY  Format   = BEST9.       _ NOPRINT   BLUE│   │
│                │   o ORDER    Spacing  = 2            _ NOZERO    RED │   │
│                │   o GROUP    Width    = 9            _ DESCENDING PINK│  │
│                │   o ACROSS   Item help =            _ PAGE     GREEN │   │
│                │   o ANALYSIS Statistic =                       CYAN │   │
│                │   * COMPUTED                                 YELLOW │   │
│                │              Type of data      Justification  WHITE │   │
│                │                * Numeric         o LEFT      ORANGE │   │
│                │                o Character       * RIGHT     BLACK  │   │
│                │                                  o CENTER    MAGENTA│   │
│                │                                              GRAY   │   │
│┌RKEYS──────────│   Header = BALANCE                           BROWN  │   │
││Add_above <F1  │                                                      │   │
││Break     <F5  │       <Edit Program>     <OK>      <Cancel>          │   │
││Delete    <F9  └──────────────────────────────────────────────────R──┘   │
└─────────────────────────────────────────────────────────────────────────┘
```

Whenever an item in a window is followed by an equals sign, you set the value of that item by typing in the field to the right of the equals sign. For example, set the format for BALANCE by positioning the cursor in the **Format** field and typing `dollar11.2`. Similarly, position the cursor in the **Width** field and type `11` for the width. When the format and width are set correctly, select the **OK** push button. The modified report appears in Display 3.8.

Display 3.8
Report after
Setting
Characteristics for
BALANCE

```
┌REPORT─────────────────────────────────────────────────────────────┐
│File Edit Search View Locals Globals                                │
│                                                                    │
│                         The SAS System                           1 │
│                                                                    │
│                                      Amount       Amount           │
│        Quarter  Department  Account  Budgeted     Spent    BALANCE  │
│             1   Equipment   lease    $40,000.00  $40,000.00    $0.00│
│                             maint    $10,000.00   $7,542.13 $2,457.87│
│                             purchase $40,000.00  $48,282.38 $-8,282.38│
│                             rental    $4,000.00   $3,998.87    $1.13│
│                             sets      $7,500.00   $8,342.68 $-842.68│
│                             tape      $8,000.00   $6,829.42 $1,170.58│
│                 Facilities  rent     $24,000.00  $24,000.00    $0.00│
│                             supplies  $2,750.00   $2,216.55  $533.45│
│                             utils     $5,000.00   $4,223.29  $776.71│
│                 Other       advert   $30,000.00  $32,476.98 $-2,476.98│
│                             musicfee  $3,000.00   $2,550.50  $449.50│
└────────────────────────────────────────────────────────────────────┘
┌RKEYS───────────────────────────────────────────────────────────────┐
│Add_above <F1>      Add_below <F2>      Add_left  <F3>   Add_right <F4>│
│Break     <F5>      CGrow     <F6>      CShrink   <F7>   Define    <F8>│
│Delete    <F9>      Move    <SHF F0>    Page    <SHF F1> Pageback <SHF F2>│
│                                                              R       │
└────────────────────────────────────────────────────────────────────┘
```

Changing a Column Header

By default, when you add a variable to a report, the REPORT procedure uses the variable's name, in uppercase letters, as the column header. It is easy to change the column header. For this report, change the header for the variable BALANCE to match the style of the other column headers by typing `Balance` anywhere in the space for the column header. (If you don't type over all the characters from the old header, be sure to delete them.)

The REPORT procedure doesn't justify the new header or honor the split character if you use one until you issue the REFRESH command, which is under **Edit** in the action bar at the top of the REPORT window. When you issue the REFRESH command, PROC REPORT right justifies the new column header so that the report looks like Display 3.9.

Note: If you type `Balance` exactly over `BALANCE`, the REFRESH command does not alter your report.

Display 3.9
Report with a New Column Header

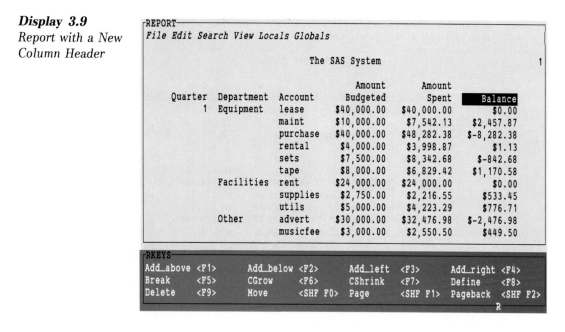

Controlling Page Breaks in a Report

By default, the location of page breaks in a report depends on the page size you use. However, you can insert page breaks in a report whenever the value of an order or group variable changes. The current report contains three order variables: QTR, DEPT, and ACCOUNT. In this section you learn how to insert a page break each time the value of QTR changes.

First select **Quarter** and issue the BREAK command, which is under **Edit** in the action bar at the top of the REPORT window. Because **Quarter** was selected when you issued the BREAK command, QTR is the *break variable*. The REPORT procedure inserts the break you request each time the value of the break variable changes.

After you issue the BREAK command, the REPORT procedure displays the LOCATION window, which appears in Display 3.10.

Display 3.10
LOCATION
Window

```
┌REPORT─────────────────────────────────────────────────────────────────────┐
│File Edit Search View Locals Globals                                        │
│                                                                            │
│                            The SAS System                                 1│
│                                                                            │
│                                     Amount       Amount                    │
│         ┌──────┐                                                           │
│         │Quarter│ Department  Account   Budgeted      Spent      Balance   │
│         └──────┘                                                           │
│             1   Equipment   lease    $40,000.00  $40,000.00       $0.00    │
│                             maint    $10,000.00   $7,542.13   $2,457.87    │
│       ┌LOCATION────────────┐chase    $40,000.00  $48,282.38  $-8,282.38    │
│       │ o After detail     │tal       $4,000.00   $3,998.87       $1.13    │
│       │ o Before detail    │s         $7,500.00   $8,342.68    $-842.68    │
│       │                    │e         $8,000.00   $6,829.42   $1,170.58    │
│       │  <OK>      <Cancel>│t        $24,000.00  $24,000.00       $0.00    │
│       └────────────────────┘plies     $2,750.00   $2,216.55     $533.45    │
│                             ls        $5,000.00   $4,223.29     $776.71    │
│              Other        advert    $30,000.00  $32,476.98  $-2,476.98    │
│                           musicfee   $3,000.00   $2,550.50     $449.50    │
│                                                                            │
└────────────────────────────────────────────────────────────────────────────┘
┌RKEYS───────────────────────────────────────────────────────────────────────┐
│Add_above <F1>     Add_below <F2>     Add_left  <F3>      Add_right <F4>     │
│Break     <F5>     CGrow     <F6>     CShrink   <F7>      Define    <F8>     │
│Delete    <F9>     Move   <SHF F0>    Page   <SHF F1>     Pageback <SHF F2>  │
│                                                                          R │
└────────────────────────────────────────────────────────────────────────────┘
```

The choices in the LOCATION window control the position of the page breaks
in the report:

After detail
    places the page break at the end of each set of rows that have the same value
    for the break variable.

Before detail
    places the page break at the beginning of each set of rows that have the same
    value for the break variable.

Select **After detail**. Next, select the **OK** push button. PROC REPORT now
displays the BREAK window, which appears in Display 3.11. Notice that the
header in this window tells you that you are creating a break after QTR.

Display 3.11
BREAK Window

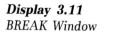

```
┌REPORT──────────────────────────────────────────────────────────────┐
│ File Edit Search View Locals Globals                                │
│                                                                     │
│                        The SAS System                             1 │
│                                                                     │
│        ┌BREAK─────────────────────┐ unt      Amount                 │
│        │     Breaking AFTER  QTR   │ ted      Spent      Balance     │
│        │ Options            Color  │.00  $40,000.00       $0.00      │
│        │ _ OVERLINE         BLUE   │.00   $7,542.13   $2,457.87      │
│        │ _ DOUBLE OVERLINE (=) RED │.00  $48,282.38  $-8,282.38      │
│        │ _ UNDERLINE        PINK   │.00   $3,998.87       $1.13      │
│        │ _ DOUBLE UNDERLINE (=) GREEN│.00  $8,342.68  $-842.68       │
│        │                    CYAN   │.00   $6,829.42   $1,170.58      │
│        │ _ SKIP             YELLOW │.00  $24,000.00       $0.00      │
│        │ _ PAGE             WHITE  │.00   $2,216.55     $533.45      │
│        │                    ORANGE │.00   $4,223.29     $776.71      │
│        │ _ SUMMARIZE        BLACK  │.00  $32,476.98  $-2,476.98      │
│        │ _ SUPPRESS         MAGENTA│.00   $2,550.50     $449.50      │
│        │                    GRAY   │                                 │
│┌RKEYS──│                    BROWN  │                                 │
││Add_a  │                          │ left  <F3>     Add_right <F4>    │
││Break  │ <Edit Program>  <OK>  <Cancel>│ ink <F7>    Define   <F8>   │
││Delet  └──────────────────────────┘      <SHF F1>  Pageback <SHF F2> │
│                                                           R         │
└─────────────────────────────────────────────────────────────────────┘
```

You will learn more about the choices in the BREAK window in later chapters. For now, just select **PAGE**. When you have done so, select the **OK** push button.

Viewing the Report

Although it may not be immediately obvious on your display, the report now starts a new page when the value of QTR changes. The amount of the first page that is visible on your display depends on the size of the display. Whether or not all the rows of the report in which QTR=1 fit on one page depends on the page size of your report. However, regardless of the size of your display and regardless of the page size, the REPORT procedure displays only one page of a report at a time.

Use the commands under **View** in the action bar at the top of the REPORT window to view the entire report.

When you have viewed the report to your satisfaction, use the necessary commands to return to the top of the first page of the report, which appears in Display 3.12.

Display 3.12
Top of the First
Page of the Report

```
┌REPORT─────────────────────────────────────────────────────────────────────┐
│File Edit Search View Locals Globals                                        │
│                                                                            │
│                            The SAS System                                 1│
│                                                                            │
│                                    Amount       Amount                     │
│          Quarter  Department  Account   Budgeted      Spent      Balance    │
│             1     Equipment   lease     $40,000.00  $40,000.00      $0.00   │
│                               maint     $10,000.00   $7,542.13  $2,457.87   │
│                               purchase  $40,000.00  $48,282.38 $-8,282.38   │
│                               rental     $4,000.00   $3,998.87      $1.13   │
│                               sets       $7,500.00   $8,342.68  $-842.68   │
│                               tape       $8,000.00   $6,829.42  $1,170.58   │
│                    Facilities rent      $24,000.00  $24,000.00      $0.00   │
│                               supplies   $2,750.00   $2,216.55    $533.45   │
│                               utils      $5,000.00   $4,223.29    $776.71   │
│                    Other      advert    $30,000.00  $32,476.98 $-2,476.98   │
│                               musicfee   $3,000.00   $2,550.50    $449.50   │
│                                                                            │
├RKEYS───────────────────────────────────────────────────────────────────────┤
│Add_above <F1>      Add_below <F2>      Add_left  <F3>      Add_right <F4>   │
│Break     <F5>      CGrow     <F6>      CShrink   <F7>      Define    <F8>   │
│Delete    <F9>      Move    <SHF F0>    Page    <SHF F1>    Pageback <SHF F2>│
│                                                                          R │
└────────────────────────────────────────────────────────────────────────────┘
```

Customizing the Title of a Report

Another way to enhance the appearance of the report is to use a title that describes the report. Use the TITLE command to change the title of the report in your display. First, select **Globals** from the action bar at the top of the REPORT window. When the pull-down menu appears, as shown in Display 3.13, select **TITLE**.

Display 3.13
Selecting the
TITLE Command

```
┌REPORT─────────────────────────────────────────────────────────────────────┐
│File Edit Search View Locals Globals                                        │
│                          ┌────────┐                                        │
│                          │Title   │  System                              1 │
│                          │Footnote│                                        │
│                          │Keys    │     Amount       Amount                │
│          Quarter Department Ac│Command│ Budgeted      Spent      Balance    │
│             1    Equipment   le└────────┘0,000.00  $40,000.00      $0.00   │
│                              maint     $10,000.00   $7,542.13  $2,457.87   │
│                              purchase  $40,000.00  $48,282.38 $-8,282.38   │
│                              rental     $4,000.00   $3,998.87      $1.13   │
│                              sets       $7,500.00   $8,342.68  $-842.68   │
│                              tape       $8,000.00   $6,829.42  $1,170.58   │
│                   Facilities rent      $24,000.00  $24,000.00      $0.00   │
│                              supplies   $2,750.00   $2,216.55    $533.45   │
│                              utils      $5,000.00   $4,223.29    $776.71   │
│                   Other      advert    $30,000.00  $32,476.98 $-2,476.98   │
│                              musicfee   $3,000.00   $2,550.50    $449.50   │
│                                                                            │
├RKEYS───────────────────────────────────────────────────────────────────────┤
│Add_above <F1>      Add_below <F2>      Add_left  <F3>      Add_right <F4>   │
│Break     <F5>      CGrow     <F6>      CShrink   <F7>      Define    <F8>   │
│Delete    <F9>      Move    <SHF F0>    Page    <SHF F1>    Pageback <SHF F2>│
│                                                                          R │
└────────────────────────────────────────────────────────────────────────────┘
```

Change the text on line 1 of the TITLE window to read `Report of Departments`. Add the text `by Quarter` on line 2. When your display looks like Display 3.14, select **Edit** from the action bar at the top of the TITLE window. Select **End** from the pull-down menu that appears.

Display 3.14
Changing the Title of the Report

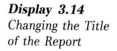

```
┌REPORT─────────────────────────────────────────────────────────────────────┐
│File Edit Search View Locals Globals                                        │
│     ┌TITLES──────────────────────────────────────────────┐               1│
│     │Edit View Globals Help                               │                │
│     │                                                     │                │
│  Qua│ Title      Value                                    │             nce│
│     │                                                     │             .00│
│     │   1        Report of Departments                    │             .87│
│     │   2        by Quarter                               │             .38│
│     │   3                                                 │             .13│
│     │   4                                                 │             .68│
│     │   5                                                 │             .58│
│     │   6                                                 │             .00│
│     │   7                                                 │             .45│
│     │   8                                                 │             .71│
│     │   9                                                 │             .98│
│     │  10                                                 │             .50│
├RKEYS┴─────────────────────────────────────────────────────────────────────┤
│Add_above <F1>      Add_below <F2>      Add_left  <F3>     Add_right <F4>    │
│Break     <F5>      CGrow     <F6>      CShrink   <F7>     Define    <F8>    │
│Delete    <F9>      Move      <SHF F0>  Page      <SHF F1> Pageback  <SHF F2>│
│                                                                           R│
└────────────────────────────────────────────────────────────────────────────┘
```

To see the report with its customized title, as shown in Display 3.15, issue the REFRESH command (under **Edit** in the action bar at the top of the REPORT window).

Display 3.15
Report with New Title

```
┌REPORT─────────────────────────────────────────────────────────────────────┐
│File Edit Search View Locals Globals                                        │
│                        Report of Departments                              1│
│                            by Quarter                                      │
│                                                                            │
│                                    Amount       Amount                     │
│     Quarter  Department  Account   Budgeted      Spent      Balance        │
│           1  Equipment   lease     $40,000.00  $40,000.00      $0.00       │
│                          maint     $10,000.00   $7,542.13  $2,457.87       │
│                          purchase  $40,000.00  $48,282.38  $-8,282.38      │
│                          rental     $4,000.00   $3,998.87      $1.13       │
│                          sets       $7,500.00   $8,342.68   $-842.68       │
│                          tape       $8,000.00   $6,829.42  $1,170.58       │
│              Facilities  rent      $24,000.00  $24,000.00      $0.00       │
│                          supplies   $2,750.00   $2,216.55    $533.45       │
│                          utils      $5,000.00   $4,223.29    $776.71       │
│              Other       advert    $30,000.00  $32,476.98  $-2,476.98      │
├RKEYS┬─────────────────────────────────────────────────────────────────────┤
│Add_above <F1>      Add_below <F2>      Add_left  <F3>     Add_right <F4>    │
│Break     <F5>      CGrow     <F6>      CShrink   <F7>     Define    <F8>    │
│Delete    <F9>      Move      <SHF F0>  Page      <SHF F1> Pageback  <SHF F2>│
│                                                                           R│
└────────────────────────────────────────────────────────────────────────────┘
```

Note: When you invoke the REPORT procedure, it honors whatever TITLE statements are in effect. Therefore, when you store a report definition, the REPORT procedure does not store the title. You can change the title during a REPORT session in order to see what the title will look like in the report, but if you store the report definition, you won't see the title when you use the definition unless you specify it during your SAS session, either in the PROC step that invokes the REPORT procedure or earlier in the session. You can simulate a permanent title with the SPAN command (see "Creating a Header That Spans Multiple Columns" in Chapter 4, "Grouping and Summarizing Data").

Separating Column Headers from the Rows of the Report

You may prefer a report with underlined column headers and a blank line between the column headers and the first row of the report. The REPORT procedure provides two options that make it easy to create such a report. The HEADLINE option underlines all column headers and the spaces between them at the top of each page of the report. The HEADSKIP option writes a blank line beneath all column headers (or beneath the underlining that HEADLINE writes) at the top of each page of the report.

To make these changes, first issue the ROPTIONS command. (This command is under **Locals** in the action bar at the top of the REPORT window.) From the list of options in the ROPTIONS window, select **HEADLINE** and **HEADSKIP**. When your display looks like Display 3.16, exit the ROPTIONS window.

Note: The line size and page size in your ROPTIONS window may differ from the line size and page size in Display 3.16. You don't need to change them.

Display 3.16
Selecting the
HEADLINE and
HEADSKIP
Options

```
┌REPORT─────────────────────────────────────────────────────────────────┐
│ File Edit Se┌ROPTIONS──────────────────────────────┐                 1 │
│             │ Modes         Attributes             │                   │
│             │ _ DEFER       Linesize   =  78       │                   │
│             │ _ PROMPT      Pagesize   =  20       │                   │
│             │               Colwidth   =   9       │                   │
│             │ Options       Spacing    =   2       │  Amount           │
│    ┌Quarter┐│ X CENTER      Split      =   /       │   Spent   Balance │
│    │     1 ││ X HEADLINE    Panels     =   1       │$40,000.00    $0.00│
│             │ X HEADSKIP    Panelspace =   4       │ $7,542.13 $2,457.87│
│             │ _ NAMED                              │$48,282.38 $-8,282.38│
│             │ _ NOHEADER         User Help         │ $3,998.87    $1.13│
│             │ _ SHOWALL     Libname =              │ $8,342.68 $-842.68│
│             │ _ WRAP        Catalog =              │ $6,829.42 $1,170.58│
│             │                                      │$24,000.00    $0.00│
│             │    <OK>        <Cancel>              │ $2,216.55  $533.45│
│             └──────────────────────────────────────┘ $4,223.29  $776.71│
│                                                      $32,476.98 $-2,476.98│
├RKEYS───────────────────────────────────────────────────────────────────┤
│ Add_above <F1>     Add_below <F2>    Add_left  <F3>    Add_right <F4>   │
│ Break     <F5>     CGrow     <F6>    CShrink   <F7>    Define    <F8>   │
│ Delete    <F9>     Move    <SHF F0>  Page    <SHF F1>  Pageback <SHF F2>│
│                                                                      R  │
└────────────────────────────────────────────────────────────────────────┘
```

Now, exit the ROPTIONS window. The modified report appears in Display 3.17.

Display 3.17
Column Headers
Separated from the
First Row of the
Report

```
┌REPORT─────────────────────────────────────────────────────────┐
│File Edit Search View Locals Globals                            │
│                                                                │
│                    Report of Departments                      1│
│                        by Quarter                              │
│                                                                │
│                            Amount      Amount                  │
│        Quarter Department  Account     Budgeted      Spent      Balance    │
│        ------------------------------------------------------------│
│                                                                │
│              1 Equipment   lease      $40,000.00  $40,000.00       $0.00 │
│                            maint      $10,000.00   $7,542.13   $2,457.87 │
│                            purchase   $40,000.00  $48,282.38  $-8,282.38 │
│                            rental      $4,000.00   $3,998.87       $1.13 │
│                            sets        $7,500.00   $8,342.68    $-842.68 │
│                            tape        $8,000.00   $6,829.42   $1,170.58 │
│                Facilities  rent       $24,000.00  $24,000.00       $0.00 │
│                            supplies    $2,750.00   $2,216.55     $533.45 │
│                                                                │
├RKEYS───────────────────────────────────────────────────────────│
│Add_above <F1>      Add_below <F2>      Add_left  <F3>    Add_right <F4> │
│Break     <F5>      CGrow     <F6>      CShrink   <F7>    Define    <F8> │
│Delete    <F9>      Move    <SHF F0>    Page    <SHF F1>  Pageback <SHF F2>│
│                                                            R   │
└────────────────────────────────────────────────────────────────┘
```

Saving the Report Definition

This report is now complete. Use the RSTORE command (under **File** in the action bar at the top of the REPORT window) to store it in SASUSER.TUTORS.CHAP3.REPT.

Using the Report Definition to Print the Report

The SAS statements that create a report with the report definition you built in this chapter follow:

```
proc report data=report.budget report=sasuser.tutors.chap3;
   title 'Report of Departments';
   title2 'by Quarter';
run;
```

After you have created the report, follow the steps described in "Printing the Report" in Chapter 2 to print a copy of the report. Remember to include TITLE statements when you invoke the procedure if you want your customized title to appear in the report.

Note: The TITLE statements are not necessary if you are still in the same SAS session in which you created the report.

Using the REPORT Language to Produce the Report

The following SAS statements produce the report you developed in this tutorial. For a detailed explanation of these statements, see "Example 2: Enhancing the Ordered Report" in Chapter 10, "The REPORT Language."

```
proc report data=report.budget headline headskip;

    title 'Report of Departments';
    title2 'by Quarter';

    column qtr dept account budget actual balance;

    define qtr      / order format=1. width=7 'Quarter';
    define dept     / order format=$10. width=10 'Department';
    define account  / order format=$8. width=8 'Account';
    define budget   / display format=dollar11.2 width=11
                        'Amount/Budgeted';
    define actual   / display format=dollar11.2 width=11
                        'Amount/Spent';
    define balance  / computed format=dollar11.2 width=11
                        'Balance';

    compute balance;
       balance=budget-actual;
    endcomp;

    break after qtr / page;

run;
```

Chapter **4** Grouping and Summarizing Data

Introduction

In this chapter you learn to use some of the more advanced features of the REPORT procedure. These features include

☐ grouping observations from the data set to create reports that summarize the data

☐ summarizing data for related rows of a report

☐ summarizing data for all rows of the report.

You also learn to create a column header that spans multiple columns and to use such a header as a title that you can store with the report.
The report you create incorporating these features appears in Output 4.1.

Output 4.1
Report with
Groups and
Summaries

```
                                                                    1

                        Year-to-Date Financial Status
                              by Department

                               Amount         Amount
          Department   Account  Budgeted       Spent        Balance
          Equipment    lease    $80,000.00   $80,000.00         $0.00
                       maint    $22,000.00   $18,217.42     $3,782.58
                       purchase $60,000.00   $66,051.53    $-6,051.53
                       rental   $10,000.00    $9,481.81       $518.19
                       sets     $15,000.00   $16,422.30    $-1,422.30
                       tape     $20,000.00   $18,256.15     $1,743.85
                                ----------   ----------    ----------
                               $207,000.00  $208,429.21    $-1,429.21
                                ----------   ----------    ----------


          Facilities   rent     $48,000.00   $48,000.00         $0.00
                       supplies  $5,500.00    $4,959.03       $540.97
                       utils     $8,500.00    $7,668.10       $831.90
                                ----------   ----------    ----------
                                $62,000.00   $60,627.13     $1,372.87
                                ----------   ----------    ----------

          Other        advert   $60,000.00   $69,802.62    $-9,802.62
                       musicfee  $8,000.00    $7,426.45       $573.55
                       talent   $33,000.00   $31,411.37     $1,588.63
                                ----------   ----------    ----------
                               $101,000.00  $108,640.44    $-7,640.44
                                ----------   ----------    ----------

          Staff        fulltime $295,000.00 $293,988.43     $1,011.57
                       parttime $100,000.00  $99,869.08       $130.92
                                ----------   ----------    ----------
                               $395,000.00  $393,857.51     $1,142.49
                                ----------   ----------    ----------

          Travel       gas       $2,000.00    $1,522.19       $477.81
                       leases    $8,000.00    $6,934.80     $1,065.20
                                ----------   ----------    ----------
                                $10,000.00    $8,456.99     $1,543.01
                                ----------   ----------    ----------

                                ==========   ==========    ==========
                               $775,000.00  $780,011.28    $-5,011.28
                                ==========   ==========    ==========
```

Note: By the time you reach this chapter, you should be comfortable issuing commands and exiting windows in the REPORT procedure. Detailed directions for doing so are no longer provided. Unless otherwise specified, commands are under **Edit** in the action bar at the top of the REPORT window.

Understanding Groups

You create groups by designating one or more variables as group variables. PROC REPORT orders the rows of a report according to the formatted values of group variables. In this sense, group variables are similar to the order variables you used in the two reports you created earlier. However, when you use group variables in conjunction with analysis variables (variables you can calculate statistics for), the REPORT procedure consolidates observations that have a unique combination of values for all group variables into one detail row of the report.

You must associate a statistic with an analysis variable. The value displayed for an analysis variable is the statistic associated with it calculated for the set of observations consolidated in that row of the report.

For example, consider Output 4.2. In this report, DEPT is a group variable, and BUDGET is an analysis variable used to calculate the SUM statistic. Each detail row of the report contains the sum of BUDGET for all observations in the data set that have a unique value for DEPT. Thus, each detail row shows the sum of BUDGET for both quarters for all accounts within one department. For example, the first detail row shows that the amount budgeted over two quarters for all equipment was $207,000.00.

Output 4.2
Creating Groups

```
                          The SAS System                            1

            DEPT              BUDGET
            Equipment    $207,000.00
            Facilities    $62,000.00
            Other        $101,000.00
            Staff        $395,000.00
            Travel        $10,000.00
```

You can use more than one group variable in a report. The report in Output 4.3 contains two group variables: DEPT and ACCOUNT. Here, BUDGET is once again an analysis variable used to calculate the SUM statistic. Each detail row of the report shows the sum of BUDGET for both quarters for all observations with a unique combination of values for DEPT and ACCOUNT. Thus, each detail row shows the sum of BUDGET for one account in one department. For example, the first detail row shows that the equipment department budgeted $80,000 for leasing over two quarters.

Output 4.3
Creating Groups
with Two Group
Variables

```
                          The SAS System                            1

        DEPT         ACCOUNT       BUDGET
        Equipment    lease      $80,000.00
                     maint      $22,000.00
                     purchase   $60,000.00
                     rental     $10,000.00
                     sets       $15,000.00
                     tape       $20,000.00
        Facilities   rent       $48,000.00
                     supplies    $5,500.00
                     utils       $8,500.00
        Other        advert     $60,000.00
                     musicfee    $8,000.00
                     talent     $33,000.00
        Staff        fulltime  $295,000.00
                     parttime  $100,000.00
        Travel       gas         $2,000.00
                     leases      $8,000.00
```

Summarizing Data

You summarize data by creating breaks in the report either when the value of a particular variable changes or at the beginning or end of the report. A break can do more than summarize the data. It is a section of the report that does one or more of the following:

□ visually separates parts of the report

□ summarizes statistics and computed variables

□ displays text, values calculated for a set of rows of the report, or both

□ executes DATA step statements (See Chapter 5, "Using Some Advanced Features of the REPORT Procedure.")

The report in Output 4.4 contains two kinds of breaks. The first kind, which occurs each time the value of DEPT changes, summarizes BUDGET, ACTUAL, and BALANCE for each department. The second kind, which appears at the end of the report, summarizes the same variables for all departments in the company.

Output 4.4
Creating Two
Kinds of Breaks

```
                                    The SAS System                          1

                                     Amount        Amount
                       Department  Account    Budgeted         Spent        Balance
                       Equipment   lease     $80,000.00    $80,000.00         $0.00
                                   maint     $22,000.00    $18,217.42     $3,782.58
                                   purchase  $60,000.00    $66,051.53    $-6,051.53
                                   rental    $10,000.00     $9,481.81       $518.19
                                   sets      $15,000.00    $16,422.30    $-1,422.30
                                   tape      $20,000.00    $18,256.15     $1,743.85
                       ----------            ----------    ----------    ----------
                       Equipment            $207,000.00   $208,429.21    $-1,429.21
                       ----------            ----------    ----------    ----------

                       Facilities  rent      $48,000.00    $48,000.00         $0.00
                                   supplies   $5,500.00     $4,959.03       $540.97
                                   utils      $8,500.00     $7,668.10       $831.90
                       ----------            ----------    ----------    ----------
                       Facilities            $62,000.00    $60,627.13     $1,372.87
                       ----------            ----------    ----------    ----------

                       Other       advert    $60,000.00    $69,802.62    $-9,802.62
                                   musicfee   $8,000.00     $7,426.45       $573.55
                                   talent    $33,000.00    $31,411.37     $1,588.63
                       ----------            ----------    ----------    ----------
                       Other                $101,000.00   $108,640.44    $-7,640.44
                       ----------            ----------    ----------    ----------

                       Staff       fulltime $295,000.00   $293,988.43     $1,011.57
                                   parttime $100,000.00    $99,869.08       $130.92
                       ----------            ----------    ----------    ----------
                       Staff                $395,000.00   $393,857.51     $1,142.49
                       ----------            ----------    ----------    ----------

                       Travel      gas        $2,000.00     $1,522.19       $477.81
                                   leases     $8,000.00     $6,934.80     $1,065.20
                       ----------            ----------    ----------    ----------
                       Travel                $10,000.00     $8,456.99     $1,543.01
                       ----------            ----------    ----------    ----------

                                            ==========    ==========    ==========
                                           $775,000.00   $780,011.28    $-5,011.28
                                            ==========    ==========    ==========
```

Invoking the REPORT Procedure

To create the report in Output 4.4, you begin with the report you completed in the previous tutorial. You have already learned how to use the REPORT= option to invoke the REPORT procedure with a stored report definition. To do so in this case, submit the following SAS statements. Because you create a new title in this tutorial, submit a null TITLE statement with the PROC REPORT statement to ensure that the initial report does not have a title.

```
proc report data=report.budget report=sasuser.tutors.chap3
            windows;
    title;
run;
```

When the REPORT procedure starts, the REPORT window contains the report in Display 4.1.

Display 4.1
Invoking PROC
REPORT with a
Stored Report
Definition

```
┌REPORT─────────────────────────────────────────────────────────────┐
│File Edit Search View Locals Globals                                │
│                                                                   1│
│                                                                    │
│                               Amount     Amount                    │
│            Quarter Department Account  Budgeted      Spent  Balance │
│            ───────────────────────────────────────────────────────│
│                                                                    │
│                1 Equipment  lease   $40,000.00  $40,000.00    $0.00 │
│                             maint   $10,000.00   $7,542.13 $2,457.87│
│                             purchase $40,000.00 $48,282.38 $-8,282.38│
│                             rental   $4,000.00   $3,998.87    $1.13 │
│                             sets     $7,500.00   $8,342.68 $-842.68 │
│                             tape     $8,000.00   $6,829.42 $1,170.58│
│                  Facilities rent    $24,000.00  $24,000.00    $0.00 │
│                             supplies $2,750.00   $2,216.55  $533.45 │
│                             utils    $5,000.00   $4,223.29  $776.71 │
└────────────────────────────────────────────────────────────────────┘
┌RKEYS───────────────────────────────────────────────────────────────┐
│Add_above <F1>    Add_below <F2>    Add_left <F3>    Add_right <F4>  │
│Break     <F5>    CGrow     <F6>    CShrink  <F7>    Define    <F8>  │
│Delete    <F9>    Move    <SHF F0>  Page   <SHF F1>  Pageback <SHF F2>│
│                                                         R           │
└────────────────────────────────────────────────────────────────────┘
```

Deleting Items in a Report

The report you are creating in this chapter does not contain the variable QTR. In addition, although the report does contain the variables ACTUAL and BALANCE, it is easier to move through the tutorial if you delete these variables now and add them again later.

You can delete an item in the report in one of several ways, depending on the number of lines that are in the column header.

Deleting an Item with a One-Line Column Header

When you delete an item with a one-line column header, you can select either the header or the entire item before you issue the DELETE command. This section illustrates both methods.

To delete QTR, select the column header **Quarter** and issue the DELETE command. When you do so, the layout of the report changes, as you can see by comparing Display 4.2 to Display 4.1. In Display 4.1 PROC REPORT orders the observations first by QTR, so all rows pertaining to the first quarter appear before any rows pertaining to the second quarter. After you delete QTR, PROC REPORT orders the rows of the report first by DEPT, then by ACCOUNT. For each combination of values for DEPT and ACCOUNT, this data set contains two observations, one for each quarter.

Display 4.2
The Structure of
the Report
Changes When You
Delete QTR

```
┌─REPORT──────────────────────────────────────────────────────────────┐
│ File Edit Search View Locals Globals                                 │
│ NOTE: QTR deleted.                                                    │
│                                                                    1 │
│                                                                      │
│                              Amount        Amount                    │
│            Department  Account  Budgeted       Spent       Balance   │
│            ──────────────────────────────────────────────────────   │
│                                                                      │
│            Equipment   lease   $40,000.00  $40,000.00        $0.00   │
│                                $40,000.00  $40,000.00        $0.00   │
│                        maint   $10,000.00   $7,542.13    $2,457.87   │
│                                $12,000.00  $10,675.29    $1,324.71   │
│                        purchase $40,000.00 $48,282.38   $-8,282.38   │
│                                $20,000.00  $17,769.15    $2,230.85   │
│                        rental   $4,000.00   $3,998.87        $1.13   │
│                                 $6,000.00   $5,482.94      $517.06   │
│                        sets     $7,500.00   $8,342.68     $-842.68   │
└──────────────────────────────────────────────────────────────────────┘
┌─RKEYS─────────────────────────────────────────────────────────────────┐
│ Add_above <F1>    Add_below <F2>    Add_left  <F3>    Add_right <F4>   │
│ Break     <F5>    CGrow     <F6>    CShrink   <F7>    Define    <F8>   │
│ Delete    <F9>    Move    <SHF F0>  Page    <SHF F1>  Pageback <SHF F2>│
│                                                                     R │
└──────────────────────────────────────────────────────────────────────┘
```

Because you deleted QTR by selecting just the column header, delete
BALANCE by selecting the entire column. To do so, select any line in the column
except the header and the lines generated by the HEADLINE and HEADSKIP
options, and issue the DELETE command.

Deleting an Item with a Multiline Column Header

If you try to delete an item with a multiline header by selecting the header, PROC
REPORT deletes only the line of the header on which you position the cursor.
This feature enables you to delete lines from a column header without deleting the
entire column. To delete an item with a multiline header, you can delete lines of
the header until only one remains. When you delete the last line of the column
header, you delete the item as well. Alternatively, you can initially select the
entire column as you did for BALANCE, and issue the DELETE command. Since
you have already deleted BALANCE by selecting the entire column, you will delete
ACTUAL by selecting the header, **Amount Spent**.

Select the first line of the header, **Amount**. The REPORT procedure highlights
the entire header, but the cursor is on the first line of the header. Now, issue the
DELETE command. As Display 4.3 shows, PROC REPORT deletes not the column,
but only the line of the header on which the cursor was positioned.

Display 4.3
*Deleting One Line
of a Column
Header*

```
┌REPORT─────────────────────────────────────────────────────────────────┐
│ File Edit Search View Locals Globals                                   │
│ NOTE: Text on header line 3 deleted.                                   │
│                                                                      1 │
│                                                                        │
│                                          Amount                        │
│                     Department  Account  Budgeted     ███Spent███      │
│                     ------------------------------------------------   │
│                                                                        │
│                     Equipment   lease    $40,000.00   $40,000.00       │
│                                          $40,000.00   $40,000.00       │
│                                 maint    $10,000.00    $7,542.13       │
│                                          $12,000.00   $10,675.29       │
│                                 purchase $40,000.00   $48,282.38       │
│                                          $20,000.00   $17,769.15       │
│                                 rental    $4,000.00    $3,998.87       │
│                                           $6,000.00    $5,482.94       │
│                                 sets      $7,500.00    $8,342.68       │
│                                                                        │
┌RKEYS───────────────────────────────────────────────────────────────────┐
│ Add_above <F1>      Add_below <F2>     Add_left  <F3>    Add_right <F4>  │
│ Break     <F5>      CGrow     <F6>     CShrink   <F7>    Define    <F8>  │
│ Delete    <F9>      Move      <SHF F0> Page      <SHF F1> Pageback <SHF F2> │
│                                                                     R   │
└────────────────────────────────────────────────────────────────────────┘
```

Note: When you delete part of a column header, PROC REPORT issues a message telling you that it deleted text on a particular line. The line number it uses in this message reflects the existence of titles, page numbers, and so forth.

Now that only one line remains in the header, you can delete the column by selecting that line and issuing the DELETE command. The new version of the report appears in Display 4.4.

Display 4.4
*Report after
Deleting QTR,
BALANCE, and
ACTUAL*

```
┌REPORT─────────────────────────────────────────────────────────────────┐
│ File Edit Search View Locals Globals                                   │
│ NOTE: ACTUAL deleted.                                                  │
│                                                                      1 │
│                                                                        │
│                                            Amount                      │
│                      ██Department██ Account  Budgeted                  │
│                      ------------------------------------              │
│                                                                        │
│                      Equipment   lease    $40,000.00                   │
│                                            $40,000.00                   │
│                                  maint    $10,000.00                    │
│                                            $12,000.00                   │
│                                  purchase $40,000.00                    │
│                                            $20,000.00                   │
│                                  rental    $4,000.00                    │
│                                            $6,000.00                    │
│                                  sets      $7,500.00                    │
│                                                                        │
┌RKEYS───────────────────────────────────────────────────────────────────┐
│ Add_above <F1>      Add_below <F2>     Add_left  <F3>    Add_right <F4>  │
│ Break     <F5>      CGrow     <F6>     CShrink   <F7>    Define    <F8>  │
│ Delete    <F9>      Move      <SHF F0> Page      <SHF F1> Pageback <SHF F2> │
│                                                                     R   │
└────────────────────────────────────────────────────────────────────────┘
```

Before continuing, it will be useful to turn off the HEADLINE and HEADSKIP options. Although you may prefer to use these options in your final report, turning them off now enables you to see more rows of the report while you are working through the tutorial. Issue the ROPTIONS command, which is under **Locals** in the action bar at the top of the REPORT window. From the ROPTIONS window, turn off the **HEADLINE** and **HEADSKIP** options by selecting them.

Note: Selecting a check box like the **HEADLINE** or **HEADSKIP** option is like throwing a switch. If the check box is already selected, selecting it again cancels the selection.

When your display looks like Display 4.5, exit the ROPTIONS window. (Remember that your values for **Linesize** and **Pagesize** may differ from those in Display 4.5.)

Display 4.5
Turning Off the
HEADLINE and
HEADSKIP
Options

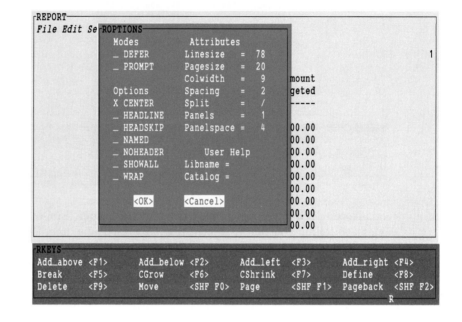

The new report appears in Display 4.6.

Display 4.6
Report without the
HEADLINE and
HEADSKIP
Options

Creating Groups

Now you are ready to create the groups in the report. To complete this process you must

□ change DEPT from an order variable to a group variable

□ change ACCOUNT from an order variable to a group variable

□ change BUDGET from a display variable to an analysis variable and assign it a statistic.

Note: The REPORT procedure cannot create groups if the report contains any order variables or any display variables that do not have a statistic above or below them. (For more information, see the description of the DEFINITION window.) Therefore, as you try to create groups in this section, the REPORT procedure displays notes until it is able to create groups.

The DEFINE command opens the DEFINITION window, through which you define the characteristics of items in the report. These characteristics include the usage, which determines how the item is used in the report. Select **Department** and issue the DEFINE command. The DEFINITION window for the variable DEPT, which appears in Display 4.7, shows that the variable is currently an order variable.

Display 4.7
DEFINITION
Window for DEPT

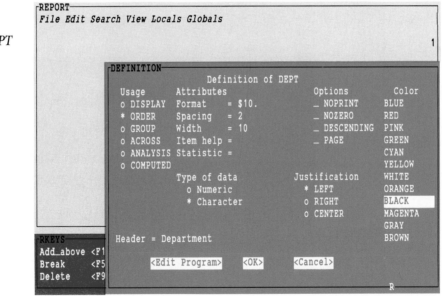

Select **GROUP** instead of **ORDER** and exit the DEFINITION window. Because PROC REPORT cannot create groups when the report contains an order variable, it issues a message telling you that it cannot create groups. This message appears in Display 4.8.

Display 4.8
Message Stating
PROC REPORT
Cannot Create
Groups because
the Report
Contains an Order
Variable

```
┌REPORT──────────────────────────────────────────────────────────────────
│File Edit Search View Locals Globals
│NOTE: Groups are not created because the usage of ACCOUNT is ORDER.
│                                                                        1
│
│                                               Amount
│                      Department  Account     Budgeted
│                      Equipment   lease       $40,000.00
│                                              $40,000.00
│                                  maint       $10,000.00
│                                              $12,000.00
│                                  purchase    $40,000.00
│                                              $20,000.00
│                                  rental       $4,000.00
│                                               $6,000.00
│                                  sets         $7,500.00
│                                               $7,500.00
│                                  tape         $8,000.00
│
├RKEYS────────────────────────────────────────────────────────────────────
│Add_above <F1>    Add_below <F2>    Add_left  <F3>      Add_right <F4>
│Break     <F5>    CGrow     <F6>    CShrink   <F7>      Define    <F8>
│Delete    <F9>    Move   <SHF F0>   Page   <SHF F1>     Pageback <SHF F2>
│                                                                        R
```

Next, select **Account** and issue the DEFINE command. The REPORT procedure opens the DEFINITION window for the variable ACCOUNT. The window appears in Display 4.9.

Display 4.9
DEFINITION
Window for
ACCOUNT

```
┌REPORT──────────────────────────────────────────────────────────────────
│File Edit Search View Locals Globals
│                                                                        1
│  ┌DEFINITION──────────────────────────────────────────────────────┐
│  │               Definition of ACCOUNT                             │
│  │  Usage      Attributes              Options         Color       │
│  │  o DISPLAY  Format   = $8.          _ NOPRINT       BLUE        │
│  │  * ORDER    Spacing  = 2            _ NOZERO        RED         │
│  │  o GROUP    Width    = 8            _ DESCENDING    PINK        │
│  │  o ACROSS   Item help =             _ PAGE          GREEN       │
│  │  o ANALYSIS Statistic =                             CYAN        │
│  │  o COMPUTED                                         YELLOW      │
│  │             Type of data          Justification     WHITE       │
│  │               o Numeric             * LEFT          ORANGE      │
│  │               * Character           o RIGHT         BLACK       │
│  │                                     o CENTER        MAGENTA     │
│  │                                                     GRAY        │
│  │  Header = Account                                   BROWN       │
│  │                                                                 │
│  │        <Edit Program>      <OK>       <Cancel>                  │
│  │                                                             R   │
```

Once again, in the DEFINITION window select **GROUP** instead of **ORDER**. Then, exit the window. PROC REPORT tells you that it cannot create groups because BUDGET is a display variable. To solve this problem, select **Amount Budgeted** and issue the DEFINE command. PROC REPORT opens the DEFINITION window for the variable BUDGET. This window appears in Display 4.10.

Display 4.10
DEFINITION
Window for
BUDGET

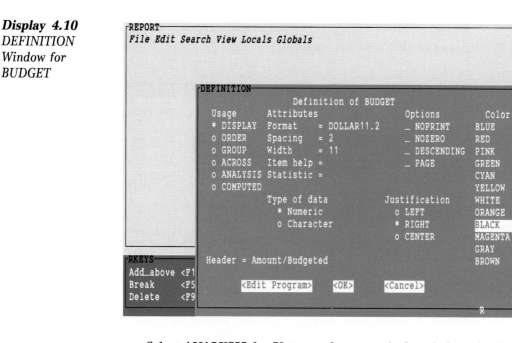

Select **ANALYSIS** for **Usage** and try to exit the window. As Display 4.11 shows, the REPORT procedure does not let you exit the DEFINITION window because you must associate a statistic with an analysis variable.

Display 4.11
Message Indicating
You Must
Associate a
Statistic with an
Analysis Variable

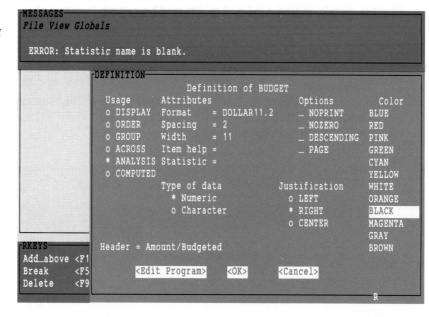

Exit or cancel the MESSAGE window. (The END and CANCEL commands are under **File** in the action bar at the top of the MESSAGE window.) Then, type **sum** in the **Statistic** field of the DEFINITION window and exit the window. The new report, which appears in Display 4.12, contains only one row for each combination of values of DEPT and ACCOUNT, whereas the previous report contained one row for each observation in the data set. Because the data set contains two observations (one for each quarter) for each account in each department, and because you defined BUDGET as an analysis variable used to calculate the SUM statistic, each value under **Amount Budgeted** is the sum of the value of BUDGET for two quarters.

Display 4.12
Using Department
and Account as
Group Variables

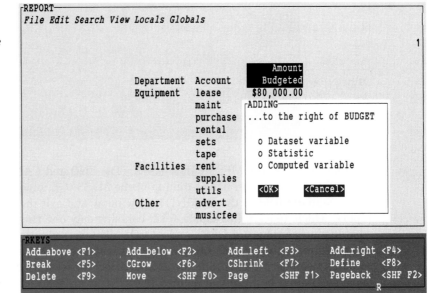

```
┌REPORT─────────────────────────────────────────────────────────────────┐
│File Edit Search View Locals Globals                                    │
│                                                                        │
│                                                                       1│
│                                                                        │
│                                                  ┌────────┐            │
│                                                  │ Amount │            │
│                               Department  Account│Budgeted│            │
│                               Equipment   lease  $80,000.00            │
│                                           maint  $22,000.00            │
│                                           purchase $60,000.00          │
│                                           rental $10,000.00            │
│                                           sets   $15,000.00            │
│                                           tape   $20,000.00            │
│                               Facilities  rent   $48,000.00            │
│                                           supplies $5,500.00           │
│                                           utils   $8,500.00            │
│                               Other       advert $60,000.00            │
│                                           musicfee $8,000.00           │
│                                                                        │
├RKEYS───────────────────────────────────────────────────────────────────┤
│Add_above <F1>      Add_below <F2>      Add_left  <F3>      Add_right <F4>│
│Break     <F5>      CGrow     <F6>      CShrink   <F7>      Define    <F8>│
│Delete    <F9>      Move      <SHF F0>  Page      <SHF F1>  Pageback <SHF F2>│
│                                                                       R│
└────────────────────────────────────────────────────────────────────────┘
```

Adding a Data Set Variable to the Report

Now that the report contains groups, add the variable ACTUAL back into the report. You want ACTUAL to appear to the right of BUDGET, so select **Amount Budgeted** and issue the ADD_RIGHT command. The ADDING window, which appears in Display 4.13, offers you three choices: **Dataset variable**, **Statistic**, and **Computed variable**. ACTUAL is a variable in the data set, so select **Dataset variable** and exit the window.

Display 4.13
ADDING a Data
Set Variable to the
Right of BUDGET

```
┌REPORT─────────────────────────────────────────────────────────────────┐
│File Edit Search View Locals Globals                                    │
│                                                                        │
│                                                                       1│
│                                                                        │
│                                                  ┌────────┐            │
│                                                  │ Amount │            │
│                               Department  Account│Budgeted│            │
│                               Equipment   lease  $80,000.00            │
│                                           maint  ┌ADDING──────────────┐│
│                                           purchase│...to the right of BUDGET│
│                                           rental  │                   ││
│                                           sets    │  o Dataset variable││
│                                           tape    │  o Statistic      ││
│                               Facilities  rent    │  o Computed variable││
│                                           supplies│                   ││
│                                           utils   │  <OK>     <Cancel>││
│                               Other       advert  │                   ││
│                                           musicfee└───────────────────┘│
│                                                                        │
├RKEYS───────────────────────────────────────────────────────────────────┤
│Add_above <F1>      Add_below <F2>      Add_left  <F3>      Add_right <F4>│
│Break     <F5>      CGrow     <F6>      CShrink   <F7>      Define    <F8>│
│Delete    <F9>      Move      <SHF F0>  Page      <SHF F1>  Pageback <SHF F2>│
│                                                                       R│
└────────────────────────────────────────────────────────────────────────┘
```

The REPORT procedure now displays the DATASET VARS window, which lists all the variables in the data set, as shown in Display 4.14.

Display 4.14
DATASET VARS
Window

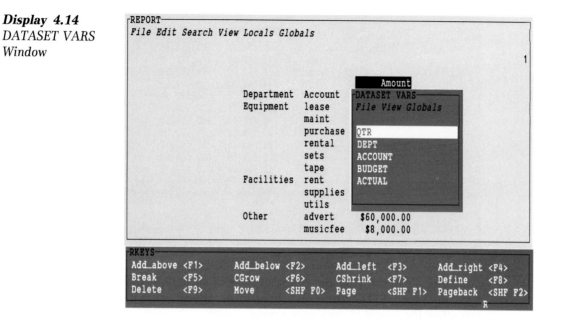

Select **ACTUAL** and exit the window. The REPORT procedure displays the DEFINITION window for **ACTUAL**, which appears in Display 4.15.

Display 4.15
DEFINITION
Window for
ACTUAL

```
┌─REPORT──────────────────────────────────────────────────────────────────┐
│ File Edit Search View Locals Globals                                     │
│                                                                        1 │
│                                                                          │
│         ┌─DEFINITION───────────────────────────────────────────────┐    │
│         │                    Definition of ACTUAL                   │    │
│         │    Usage      Attributes              Options      Color  │    │
│         │    o DISPLAY  Format     = DOLLAR11.2  _ NOPRINT    BLUE   │    │
│         │    o ORDER    Spacing    = 2           _ NOZERO     RED    │    │
│         │    o GROUP    Width      = 11          _ DESCENDING PINK   │    │
│         │    o ACROSS   Item help  =             _ PAGE       GREEN  │    │
│         │    * ANALYSIS Statistic  = SUM                      CYAN   │    │
│         │    o COMPUTED                                       YELLOW │    │
│         │               Type of data        Justification    WHITE  │    │
│         │                 * Numeric           o LEFT          ORANGE │    │
│         │                 o Character         * RIGHT         BLACK  │    │
│         │                                     o CENTER        MAGENTA│    │
│         │                                                     GRAY   │    │
│ ┌─RKEYS─│    Header = ACTUAL                                  BROWN  │    │
│ │Add_above <F1│                                                     │    │
│ │Break    <F5 │   <Edit Program>    <OK>      <Cancel>              │    │
│ │Delete   <F9 │                                                     │    │
│         └──────────────────────────────────────────────────────R───┘    │
└──────────────────────────────────────────────────────────────────────────┘
```

In this window, accept the default usage for ACTUAL. By default, PROC REPORT uses numeric variables as analysis variables used to calculate the SUM statistic. (BUDGET was a display variable rather than an analysis variable because you defined it that way when you created your first report with the PROMPT facility and had not changed its usage since then.) When you exit the window, the REPORT procedure displays the new version of the report, which appears in Display 4.16. Each value under **ACTUAL** is the sum of ACTUAL for two quarters.

Display 4.16
Report with
ACTUAL Added

```
┌REPORT─────────────────────────────────────────────────────────────────
│ File Edit Search View Locals Globals
│                                                                       1
│
│                                          ┌───────┐
│                                          │Amount │
│                   Department  Account    │Budgeted│        ACTUAL
│                   Equipment   lease       $80,000.00    $80,000.00
│                               maint       $22,000.00    $18,217.42
│                               purchase    $60,000.00    $66,051.53
│                               rental      $10,000.00     $9,481.81
│                               sets        $15,000.00    $16,422.30
│                               tape        $20,000.00    $18,256.15
│                   Facilities  rent        $48,000.00    $48,000.00
│                               supplies     $5,500.00     $4,959.03
│                               utils        $8,500.00     $7,668.10
│                   Other       advert      $60,000.00    $69,802.62
│                               musicfee     $8,000.00     $7,426.45
│
├RKEYS──────────────────────────────────────────────────────────────────
│ Add_above <F1>      Add_below <F2>      Add_left° <F3>      Add_right <F4>
│ Break     <F5>      CGrow     <F6>      CShrink   <F7>      Define    <F8>
│ Delete    <F9>      Move     <SHF F0>   Page     <SHF F1>   Pageback <SHF F2>
│                                                                       R
```

Note: You can alter the characteristics of an item in the report at any time by issuing the DEFINE command. Remember, though, that the usage characteristic can affect the layout of the entire report.

Computing a Variable Based on Statistics

In this section you add the variable BALANCE back to the report. Recall that in the previous report you did so by assigning it a value of BUDGET−ACTUAL in the COMPUTE window. If you were to try the same thing now, the column for BALANCE would contain nothing but missing values because the values in the report are no longer simply values of BUDGET and ACTUAL; they are the sums of these values for sets of observations that have the same values of the two group variables, DEPT and ACCOUNT. In this section you learn how to calculate a variable based on statistics.

To add BALANCE to the right of ACTUAL, select **ACTUAL** and issue the ADD_RIGHT command. From the ADDING window, select **Computed variable,** as shown in Display 4.17.

Display 4.17
Selecting
Computed variable
from the ADDING
Window

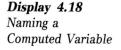

```
┌REPORT─────────────────────────────────────────────────────────────────┐
│ File Edit Search View Locals Globals                                   │
│                                                                      1 │
│                                                                        │
│                                          Amount                        │
│                  Department  Account    Budgeted      ACTUAL           │
│                  Equipment   lease     $80,000.00   $80,000.00         │
│                              maint     $22,000.00   $18,217.42         │
│                              purchase  $60,000.00┌ADDING──────────────┐│
│                              rental    $10,000.00│...to the right of ACTUAL│
│                              sets      $15,000.00│                    ││
│                              tape      $20,000.00│ o Dataset variable ││
│                  Facilities  rent      $48,000.00│ o Statistic        ││
│                              supplies   $5,500.00│ * Computed variable││
│                              utils      $8,500.00│                    ││
│                  Other       advert    $60,000.00│ <OK>      <Cancel> ││
│                              musicfee   $8,000.00└────────────────────┘│
│                                                                        │
├RKEYS───────────────────────────────────────────────────────────────────┤
│ Add_above <F1>      Add_below <F2>     Add_left  <F3>    Add_right <F4> │
│ Break     <F5>      CGrow     <F6>     CShrink   <F7>    Define    <F8> │
│ Delete    <F9>      Move   <SHF F0>    Page   <SHF F1>   Pageback <SHF F2>│
│                                                                      R │
└────────────────────────────────────────────────────────────────────────┘
```

When you exit the ADDING window, the REPORT procedure displays the COMPUTED VAR window. Type `balance` in the **Variable name** field. Because this variable is numeric, you can ignore the **Character data** check box and the **Length** prompt. You will use those features in Chapter 5. When your display looks like Display 4.18, select the **Edit Program** push button.

Note: If you inadvertently select the **OK** push button instead of the **Edit Program** push button, you return to the REPORT window. Your report contains a column header for BALANCE, but no values. To get back to where you should be, select **BALANCE**, issue the DEFINE command, and from the DEFINITION window select the **Edit Program** push button.

Display 4.18
Naming a
Computed Variable

```
┌REPORT─────────────────────────────────────────────────────────────────┐
│ File Edit Search View Locals Globals                                   │
│                                                                      1 │
│                                                                        │
│                                          Amount                        │
│                  Department  Account    Budgeted      ACTUAL           │
│                  Equipment   lease     $80,000.00   $80,000.00         │
│                              maint     $22,000.00   $18,217.42         │
│                              purchase  $6┌COMPUTED VAR───────────────┐ │
│                              rental    $1│                           │ │
│                              sets      $1│  Variable name: balance   │ │
│                              tape      $2│   _ Character data        │ │
│                  Facilities  rent      $4│    Length =               │ │
│                              supplies   $│                           │ │
│                              utils      $│ <Edit Program>  <OK>  <Cancel>│
│                  Other       advert    $6│                           │ │
│                              musicfee   $└───────────────────────────┘ │
│                                                                        │
├RKEYS───────────────────────────────────────────────────────────────────┤
│ Add_above <F1>      Add_below <F2>     Add_left  <F3>    Add_right <F4> │
│ Break     <F5>      CGrow     <F6>     CShrink   <F7>    Define    <F8> │
│ Delete    <F9>      Move   <SHF F0>    Page   <SHF F1>   Pageback <SHF F2>│
│                                                                      R │
└────────────────────────────────────────────────────────────────────────┘
```

After you select the **Edit Program** push button, PROC REPORT displays the COMPUTE window.

When referring to an analysis variable in calculations in the COMPUTE window, you use a compound name that identifies both the original variable and the statistic that the REPORT procedure now calculates from it. The compound name has the following form:

variable-name.statistic

The variables BUDGET and ACTUAL are currently analysis variables used to calculate the SUM statistic. Therefore, any statements you use in the COMPUTE window must refer to these variables by compound names. The assignment statement in Display 4.19 uses compound names to define BALANCE. Enter this statement in your COMPUTE window and exit the window.

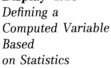

Display 4.19
Defining a
Computed Variable
Based
on Statistics

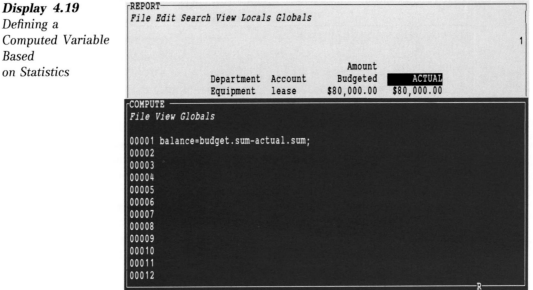

When you exit the COMPUTE window, the REPORT procedure displays the COMPUTED VAR window. Exit that window as well. The new report, which contains a column for BALANCE, appears in Display 4.20.

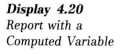

Display 4.20
*Report with a
Computed Variable*

```
┌REPORT─────────────────────────────────────────────────────────────────┐
│File Edit Search View Locals Globals                                    │
│                                                                       1│
│                                                                        │
│                                   Amount                               │
│                  Department  Account    Budgeted    ACTUAL    BALANCE   │
│                  Equipment   lease     $80,000.00  $80,000.00       0   │
│                              maint     $22,000.00  $18,217.42    3783   │
│                              purchase  $60,000.00  $66,051.53   -6052   │
│                              rental    $10,000.00   $9,481.81     518   │
│                              sets      $15,000.00  $16,422.30   -1422   │
│                              tape      $20,000.00  $18,256.15    1744   │
│                  Facilities  rent      $48,000.00  $48,000.00       0   │
│                              supplies   $5,500.00   $4,959.03     541   │
│                              utils      $8,500.00   $7,668.10     832   │
│                  Other       advert    $60,000.00  $69,802.62   -9803   │
│                              musicfee   $8,000.00   $7,426.45     574   │
│                                                                        │
├RKEYS───────────────────────────────────────────────────────────────────┤
│Add_above <F1>     Add_below <F2>     Add_left  <F3>    Add_right <F4>   │
│Break     <F5>     CGrow     <F6>     CShrink   <F7>    Define    <F8>   │
│Delete    <F9>     Move    <SHF F0>   Page    <SHF F1>  Pageback <SHF F2>│
│                                                                      R  │
└────────────────────────────────────────────────────────────────────────┘
```

By default, when you add a numeric computed variable to a report, the
REPORT procedure uses the default column width (9 unless you change it in the
ROPTIONS window) and the BEST*w.* format that spans the column. For instance,
in this case the definition of BALANCE includes a width of 9 (the default) and a
format of BEST9. To change the format and width to match those of the other
numeric variables in this report, select **BALANCE** and issue the DEFINE
command, which opens the DEFINITION window for BALANCE, as shown in
Display 4.21.

Display 4.21
*DEFINITION
Window for
BALANCE*

```
┌REPORT─────────────────────────────────────────────────────────────────┐
│File Edit Search View Locals Globals                                    │
│                                                                       1│
│       ┌DEFINITION───────────────────────────────────────────────┐     │
│   Dep │              Definition of BALANCE                       │     │
│   Equ │  Usage      Attributes              Options       Color  │     │
│       │  o DISPLAY  Format   = BEST9.       _ NOPRINT     BLUE   │     │
│       │  o ORDER    Spacing  = 2            _ NOZERO      RED    │     │
│       │  o GROUP    Width    = 9            _ DESCENDING  PINK   │     │
│       │  o ACROSS   Item help =             _ PAGE        GREEN  │     │
│       │  o ANALYSIS Statistic =                            CYAN   │     │
│   Fac │  * COMPUTED                                        YELLOW │     │
│       │             Type of data        Justification     WHITE  │     │
│       │                * Numeric          o LEFT          ORANGE │     │
│   Oth │                o Character        * RIGHT         BLACK  │     │
│       │                                   o CENTER        MAGENTA│     │
│       │                                                   GRAY   │     │
│ RKEYS │  Header = BALANCE                                  BROWN  │     │
│Add_above<F1                                                      │     │
│Break    <F5   <Edit Program>    <OK>      <Cancel>               │     │
│Delete   <F9                                                      │     │
│       └──────────────────────────────────────────────────────────┘  R │
└────────────────────────────────────────────────────────────────────────┘
```

Notice that the usage for BALANCE is computed. You cannot change the usage
for a computed variable.

Type **DOLLAR11.2** in the **Format** field and **11** in the **Width** field in the
DEFINITION window. Then exit the window. The new version of the report,

which appears in Display 4.22, displays BALANCE in the same format and in the same sized column as it displays BUDGET and ACTUAL.

Display 4.22
Altering the
Characteristics of
BALANCE to
Match Other
Numeric Variables
in the Report

```
┌REPORT─────────────────────────────────────────────────────────┐
│File Edit Search View Locals Globals                            │
│                                                              1 │
│                                                                │
│                                 Amount                         │
│            Department  Account  Budgeted       ACTUAL  BALANCE │
│            Equipment   lease   $80,000.00  $80,000.00    $0.00 │
│                        maint   $22,000.00  $18,217.42 $3,782.58│
│                        purchase $60,000.00 $66,051.53 $-6,051.53│
│                        rental  $10,000.00   $9,481.81  $518.19 │
│                        sets    $15,000.00  $16,422.30 $-1,422.30│
│                        tape    $20,000.00  $18,256.15 $1,743.85│
│            Facilities  rent    $48,000.00  $48,000.00    $0.00 │
│                        supplies $5,500.00   $4,959.03  $540.97 │
│                        utils    $8,500.00   $7,668.10  $831.90 │
│            Other       advert  $60,000.00  $69,802.62 $-9,802.62│
│                        musicfee $8,000.00   $7,426.45  $573.55 │
│                                                                │
├RKEYS───────────────────────────────────────────────────────────┤
│Add_above <F1>     Add_below <F2>    Add_left  <F3>   Add_right <F4>│
│Break     <F5>     CGrow     <F6>    CShrink   <F7>   Define    <F8>│
│Delete    <F9>     Move      <SHF F0> Page     <SHF F1> Pageback <SHF F2>│
│                                                              R │
└────────────────────────────────────────────────────────────────┘
```

Adding Break Lines for Each Department

In many cases it is useful to visually separate and summarize certain parts of a report. For instance, this report would be easier to read if it contained some kind of visual break between departments. In addition, it would be more informative if it summarized the data for each department. With PROC REPORT, you can easily add both features to the report.

You can put a break in a report whenever the value of a group or order variable changes. This group or order variable is the *break variable*. In this case, DEPT is the desired break variable.

Select **Department** and issue the BREAK command. The BREAK command opens the LOCATION window, which appears in Display 4.23. In the LOCATION window you specify whether you want the break to occur before or after the detail lines for each department.

Display 4.23
LOCATION
Window

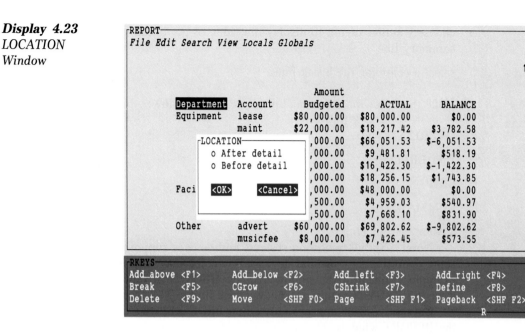

In the report you are creating, the visual breaks and the summaries occur after the last row for a department, so select **After detail** and exit the LOCATION window. The REPORT procedure now displays the BREAK window, which appears in Display 4.24.

Display 4.24
BREAK Window

The BREAK window enables you to

□ select how to visually separate the departments from each other

□ specify that you want to summarize statistics and computed variables for each department

□ choose whether or not to display the value of the break variable in the summary line

□ select a color for the break lines.

Make the following selections in the BREAK window:

□ **OVERLINE**

□ **UNDERLINE**

□ **SKIP**

□ **SUMMARIZE**

□ a color of your choice.
> **Note:** Not all operating systems and devices support all colors, and on some operating systems and devices, one color may map to another color. For example, if the BREAK window displays **BROWN** in yellow characters, selecting **BROWN** results in yellow break lines.
>
> Currently, color appears only when PROC REPORT displays the report in the REPORT window.

Next, select the **OK** push button to exit the BREAK window. As a result of your choices in the BREAK window, the REPORT procedure

□ sets the summary lines off from the rest of the report with a row of overlining and a row of underlining.

□ skips a line after each set of break lines.

□ displays a summary line for each department. The summary line contains the value of the break variable, DEPT, as well as values for BUDGET.SUM, ACTUAL.SUM, and BALANCE for all accounts in the department.

□ writes all the break lines in the color you selected.

This version of the report appears in Display 4.25, which shows the first set of break lines.

Display 4.25
Adding a Break
after DEPT

```
┌REPORT─────────────────────────────────────────────────────────────┐
│File Edit Search View Locals Globals                                │
│                                                                    │
│                                                                   1│
│                                                                    │
│                                Amount                              │
│                 Department  Account   Budgeted      ACTUAL    BALANCE│
│                 Equipment   lease   $80,000.00  $80,000.00     $0.00│
│                             maint   $22,000.00  $18,217.42 $3,782.58│
│                             purchase $60,000.00  $66,051.53 $-6,051.53│
│                             rental  $10,000.00   $9,481.81   $518.19│
│                             sets    $15,000.00  $16,422.30 $-1,422.30│
│                             tape    $20,000.00  $18,256.15 $1,743.85│
│                 ----------          -----------  ----------- -----------│
│                 Equipment           $207,000.00 $208,429.21 $-1,429.21│
│                 ----------          -----------  ----------- -----------│
│                                                                    │
│                 Facilities  rent    $48,000.00  $48,000.00     $0.00│
└────────────────────────────────────────────────────────────────────┘
┌RKEYS───────────────────────────────────────────────────────────────┐
│Add_above <F1>      Add_below <F2>      Add_left  <F3>    Add_right <F4>│
│Break     <F5>      CGrow     <F6>      CShrink   <F7>    Define    <F8>│
│Delete    <F9>      Move      <SHF F0>  Page  <SHF F1>  Pageback <SHF F2>│
│                                                              R      │
└────────────────────────────────────────────────────────────────────┘
```

Use the paging and scrolling commands under **View** in the action bar at the top of the REPORT window to browse through the report and view all the break lines. When you finish viewing the report, return to the top left corner of the first page of the report.

Adding Break Lines for the Whole Report

In addition to writing break lines for each department, you can write break lines that display a summary of the statistics and computed variables for the whole report. This process is similar to writing break lines for each department, but because you are working with the entire report, you don't need to select a break variable. In this section you write break lines that summarize the statistics and computed variables for all departments in the company.

Issue the RBREAK command, which is under **Edit** in the action bar at the top of the REPORT window. (Because a report break does not use a break variable, you can issue the RBREAK command regardless of what item is currently selected.) When the LOCATION window appears, select **After detail** to indicate that you want the break lines to appear at the end of the report. Then, exit the window.

When PROC REPORT displays the BREAK window, select

□ **DOUBLE OVERLINE**

□ **DOUBLE UNDERLINE**

□ **SUMMARIZE**

□ a color of your choice.

When you exit the BREAK window, the REPORT procedure displays the top of the first page of the report, which doesn't show the break you just put in. Use the PAGE command (under **View** in the action bar at the top of the REPORT window) repeatedly to get to the last page of the report. If necessary, scroll to the bottom of the page. The summary for the entire report appears in Display 4.26.

Display 4.26
Adding a Break at
the End of the
Report

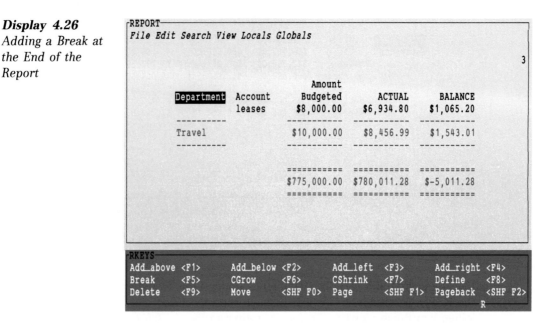

The break lines at the end of the report are similar to those you added at the breaks for each department. However, here, the summary line contains values for BUDGET.SUM, ACTUAL.SUM, and BALANCE for the entire report. You can see that for the entire company

□   the amount budgeted was $775,000.00

□   the amount spent was $780,011.28

□   the amount spent exceeded the amount budgeted by $5,011.28.

Note: If you had associated a different statistic with BUDGET and ACTUAL, the values for these variables in the summary line would be the value of that statistic calculated for the variables for all rows of the report. For instance, if you had selected the MEAN statistic, the values in the summary line would be the mean of BUDGET and ACTUAL calculated for the whole report.

Adjusting Column Headers

Use the PAGEBACK command to return to the first page of the report. Change the column header 'BALANCE' to 'Balance' to match the style of the rest of the report by typing **Balance** anywhere in the space for the column header. (If you don't type over all the characters from the old header, be sure to delete them.) If necessary, issue the REFRESH command to justify the header.

To change the column header 'ACTUAL' to 'Amount Spent', first type **Amount** over **ACTUAL**. Then select the column header and issue the ADD_BELOW command. The REPORT procedure displays the ADDING window, which appears in Display 4.27.

Display 4.27
ADDING Window
for ADD_BELOW

```
┌REPORT──────────────────────────────────────────────────────────────┐
│ File Edit Search View Locals Globals                                │
│                                                                   1 │
│                              Amount     ▐▀▀▀▀▀▀▀▀▌                   │
│             Department  Account   Budgeted ▐ Amount ▌    Balance    │
│             Equipment   lease    $80,000.00 $80,000.00     $0.00    │
│                         maint    $22,000.00 $18,217.42 $3,782.58    │
│                         purchase $60,000.00  $66,0┌ADDING─────────┐ │
│                         rental   $10,000.00   $9,4│...below ACTUAL│ │
│                         sets     $15,000.00  $16,4│               │ │
│                         tape     $20,000.00  $18,2│ o Dataset variable│
│                                                    │ o Statistic   │ │
│             ─────────            ─────────── ──────│ o Header line │ │
│             Equipment            $207,000.00 $208,4│               │ │
│             ─────────            ─────────── ──────│ <OK>   <Cancel>│ │
│                                                    └───────────────┘ │
│             Facilities  rent     $48,000.00  $48,0                   │
└─────────────────────────────────────────────────────────────────────┘
┌RKEYS─────────────────────────────────────────────────────────────┐
│ Add_above <F1>      Add_below <F2>     Add_left  <F3>   Add_right <F4>│
│ Break     <F5>      CGrow     <F6>     CShrink   <F7>   Define    <F8>│
│ Delete    <F9>      Move      <SHF F0> Page      <SHF F1> Pageback <SHF F2>│
│                                                                    R │
└─────────────────────────────────────────────────────────────────────┘
```

Notice the header in the window that tells you that you are adding below ACTUAL. Also, notice that when you open the ADDING window with the ADD_BELOW (or ADD_ABOVE) command, the third choice in the window is **Header line**, not **Computed variable**.

From the ADDING window, select **Header line** and exit the window. The REPORT procedure adds a blank line to the column header. Move the cursor to the new line and type `Spent`. Issue the REFRESH command to right-justify each line of each column header. The new column headers appear in Display 4.28.

Display 4.28
Changing Column
Headers

```
┌REPORT──────────────────────────────────────────────────────────────┐
│ File Edit Search View Locals Globals                                │
│                                                                   1 │
│                              Amount     ▐▀▀▀▀▀▀▀▀▌                   │
│                                         ▐ Amount ▌                   │
│             Department  Account   Budgeted ▐  Spent ▌    Balance    │
│             Equipment   lease    $80,000.00 $80,000.00     $0.00    │
│                         maint    $22,000.00 $18,217.42 $3,782.58    │
│                         purchase $60,000.00 $66,051.53 $-6,051.53    │
│                         rental   $10,000.00  $9,481.81   $518.19    │
│                         sets     $15,000.00 $16,422.30 $-1,422.30    │
│                         tape     $20,000.00 $18,256.15 $1,743.85    │
│                                                                     │
│             ─────────            ─────────── ─────────── ───────────│
│             Equipment            $207,000.00 $208,429.21 $-1,429.21  │
│             ─────────            ─────────── ─────────── ───────────│
│                                                                     │
│             Facilities  rent     $48,000.00 $48,000.00     $0.00    │
└─────────────────────────────────────────────────────────────────────┘
┌RKEYS─────────────────────────────────────────────────────────────┐
│ Add_above <F1>      Add_below <F2>     Add_left  <F3>   Add_right <F4>│
│ Break     <F5>      CGrow     <F6>     CShrink   <F7>   Define    <F8>│
│ Delete    <F9>      Move      <SHF F0> Page      <SHF F1> Pageback <SHF F2>│
│                                                                    R │
└─────────────────────────────────────────────────────────────────────┘
```

Note: You can also change column headers through the DEFINITION window. You will see how to do so in Chapter 5.

Creating a Header That Spans Multiple Columns

The REPORT procedure provides a command, the SPAN command, that enables you to create a header that spans multiple columns. You can use this feature to create a header that looks like a title, that is, a header that spans all columns of the report. The RSTORE command stores this header with your report definition. To use this header as a title, issue a null TITLE statement (TITLE;) before you invoke PROC REPORT.

To create a header that spans multiple columns, you must do the following:

1. Select the leftmost variable that you want the header to cover.

2. Issue the SPAN command.

3. Select the rightmost variable that you want the header to cover.

4. Issue the ADD_ABOVE command.

5. Enter the text for the header. If the header needs more lines, use the ADD_ABOVE or the ADD_BELOW command to add empty header lines. Then type in the text you want.

The title you want to add is two lines of text that you want centered over the whole report, from the leftmost column, **Department**, to the rightmost column, **Balance**. A blank line separates the text of the header from the rest of the report.

Select **Department** and issue the SPAN command, which is under **Edit** in the action bar at the top of the REPORT window. The REPORT procedure writes a message to the REPORT window telling you how to proceed. This message appears in Display 4.29.

Display 4.29
Message Indicating the Left Bound of a Spanning Header

```
┌REPORT─────────────────────────────────────────────────────────────────┐
│ File Edit Search View Locals Globals                                   │
│ NOTE: Left bound is DEPT. Select right bound then ADD_ABOVE.           │
│                                                                     1  │
│                                                                        │
│                                     Amount      Amount                 │
│              Department  Account   Budgeted      Spent      Balance    │
│              Equipment   lease    $80,000.00  $80,000.00       $0.00   │
│                          maint    $22,000.00  $18,217.42   $3,782.58   │
│                          purchase $60,000.00  $66,051.53  $-6,051.53   │
│                          rental   $10,000.00   $9,481.81     $518.19   │
│                          sets     $15,000.00  $16,422.30  $-1,422.30   │
│                          tape     $20,000.00  $18,256.15   $1,743.85   │
│                                                                        │
│              ----------           ----------  ----------  ----------   │
│              Equipment           $207,000.00 $208,429.21  $-1,429.21   │
│              ----------           ----------  ----------  ----------   │
│                                                                        │
│              Facilities  rent     $48,000.00  $48,000.00       $0.00   │
└────────────────────────────────────────────────────────────────────────┘
┌RKEYS───────────────────────────────────────────────────────────────────┐
│ Add_above <F1>      Add_below <F2>      Add_left  <F3>    Add_right <F4> │
│ Break     <F5>      CGrow     <F6>      CShrink   <F7>    Define    <F8> │
│ Delete    <F9>      Move      <SHF F0>  Page      <SHF F1> Pageback <SHF F2> │
│                                                                     R  │
└────────────────────────────────────────────────────────────────────────┘
```

Next, select **Balance** and issue the ADD_ABOVE command. PROC REPORT displays and highlights a blank column header that spans the entire report, as shown in Display 4.30.

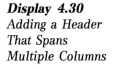

Display 4.30
*Adding a Header
That Spans
Multiple Columns*

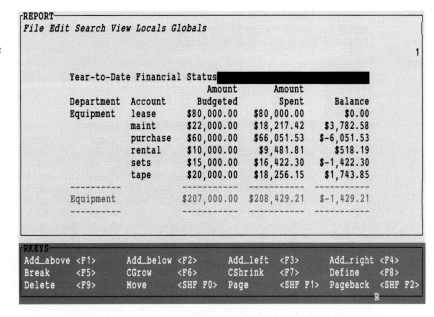

Type **Year-to-Date Financial Status** on this blank line, as shown in
Display 4.31. Don't worry about centering the text. PROC REPORT centers it for
you when it refreshes the display.

Display 4.31
*Adding Text to a
Header That Spans
Multiple Columns*

Next, issue the ADD_BELOW command to open the ADDING window. Select
Header line from the ADDING window. When your display looks like Display
4.32, exit the ADDING window.

Display 4.32
Adding Lines to
the Header

```
┌REPORT─────────────────────────────────────────────────────────┐
│ File Edit Search View Locals Globals                           │
│                                                             1  │
│                                                                │
│        Year-to-Date Financial Status███████████               │
│                                 Amount      Amount             │
│          Department  Account    Budgeted    Spent    Balance   │
│          Equipment   lease      $80,0┌ADDING──────────────┐00  │
│                      maint      $22,0│...below (          │58  │
│                      purchase   $60,0│                    │53  │
│                      rental     $10,0│  o Dataset variable│19  │
│                      sets       $15,0│  o Statistic       │30  │
│                      tape       $20,0│  * Header line     │85  │
│                                      │                    │--  │
│          Equipment              $207,0│ <OK>      <Cancel> │21  │
│                                      └────────────────────┘--  │
│                                                                │
├RKEYS───────────────────────────────────────────────────────────┤
│ Add_above <F1>   Add_below <F2>   Add_left  <F3>    Add_right <F4> │
│ Break     <F5>   CGrow     <F6>   CShrink   <F7>    Define    <F8> │
│ Delete    <F9>   Move    <SHF F0> Page    <SHF F1>  Pageback <SHF F2>│
│                                                             R  │
└────────────────────────────────────────────────────────────────┘
```

Repeat this process to add a second blank line to the header. Position the cursor on the first blank line after `Year-to-Date Financial Status` and type `Grouped by Department`. Issue the REFRESH command to center the new column header. The modified report appears in Display 4.33.

Display 4.33
Report with
Header Spanning
Multiple Columns
and Serving as
Title

```
┌REPORT─────────────────────────────────────────────────────────┐
│ File Edit Search View Locals Globals                           │
│                                                             1  │
│                                                                │
│          ┌──────────────────────────────────────┐             │
│          │      Year-to-Date Financial Status    │             │
│          │         Grouped by Department         │             │
│          └──────────────────────────────────────┘             │
│                                 Amount      Amount             │
│          Department  Account    Budgeted    Spent      Balance │
│          Equipment   lease      $80,000.00  $80,000.00    $0.00 │
│                      maint      $22,000.00  $18,217.42 $3,782.58│
│                      purchase   $60,000.00  $66,051.53 $-6,051.53│
│                      rental     $10,000.00   $9,481.81   $518.19│
│                      sets       $15,000.00  $16,422.30 $-1,422.30│
│                      tape       $20,000.00  $18,256.15 $1,743.85│
│                                 ----------  ----------  ----------│
│          Equipment              $207,000.00 $208,429.21 $-1,429.21│
│                                                                │
├RKEYS───────────────────────────────────────────────────────────┤
│ Add_above <F1>   Add_below <F2>   Add_left  <F3>    Add_right <F4> │
│ Break     <F5>   CGrow     <F6>   CShrink   <F7>    Define    <F8> │
│ Delete    <F9>   Move    <SHF F0> Page    <SHF F1>  Pageback <SHF F2>│
│                                                             R  │
└────────────────────────────────────────────────────────────────┘
```

Note: If your report contains multiple panels, you cannot use a column header that spans multiple columns as a title for the report because the header appears above each panel. For information on panels see the documentation for the ROPTIONS window.

Suppressing Parts of the Break Lines

By default, when you use the REPORT procedure to create a break on a break variable and you select underlining or overlining, the underlining or overlining appears in all columns of the report for which PROC REPORT can display a value in a summary line. In addition, summary lines contain the value of the break variable. You can easily alter the report so that the value of the break variable does not appear in the summary line. When you suppress the value of the break variable in the summary line, you also suppress underlining and overlining in the break lines in the column containing the break variable.

To make this change to your report, you must modify the break after DEPT, this time selecting **SUPPRESS** in the BREAK window.

First, select DEPT as the break variable and issue the BREAK command. Select **After detail** from the LOCATION window. Leave all selections in the BREAK window as you have them, but in addition, select **SUPPRESS**. When your display looks like Display 4.34, exit the BREAK window.

Display 4.34
Modifying the
BREAK Window

```
┌REPORT─────────────────────────────────────────────────────────────┐
│ File Edit Search View Locals Globals                               │
│                                                                  1 │
│                                                                    │
│          ┌BREAK──────────────────────────────────┐s                │
│          │          Breaking AFTER  DEPT          │                │
│          │ Options                     Color      │                │
│          │   X OVERLINE                BLUE       │unt             │
│     Depa │   _ DOUBLE OVERLINE  (=)    RED        │ent    Balance  │
│     Equi │   X UNDERLINE               PINK       │.00      $0.00  │
│          │   _ DOUBLE UNDERLINE (=)    GREEN      │.42  $3,782.58  │
│          │                             CYAN       │.53 $-6,051.53  │
│          │   X SKIP                    YELLOW     │.81    $518.19  │
│          │   _ PAGE                    WHITE      │.30 $-1,422.30  │
│          │                             ORANGE     │.15  $1,743.85  │
│          │   X SUMMARIZE               BLACK      │---  ---------- │
│     Equi │   X SUPPRESS                MAGENTA    │.21 $-1,429.21  │
│          │                             GRAY       │                │
│┌RKEYS──  │                             BROWN      │                │
│Add_above │                                       │3>    Add_right <F4>
│Break     │  <Edit Program>    <OK>    <Cancel>   │7>    Define    <F8>
│Delete    └───────────────────────────────────────┘HF F1> Pageback <SHF F2>
│                                                                 R  │
└────────────────────────────────────────────────────────────────────┘
```

The first break in the final version of the report appears in Display 4.35. The value of the break variable no longer appears in the summary line. Notice that the REPORT procedure writes underlining and overlining characters only in the columns containing values in the summary line.

Display 4.35
Suppressing the
Display of the
Value of the Break
Variable

```
┌REPORT────────────────────────────────────────────────────────────────┐
│File Edit Search View Locals Globals                                   │
│                                                                      1│
│                                                                       │
│                       Year-to-Date Financial Status                   │
│                          Grouped by Department                        │
│                                                                       │
│                             Amount      Amount                        │
│               Department  Account    Budgeted      Spent     Balance  │
│               Equipment   lease     $80,000.00  $80,000.00      $0.00  │
│                           maint     $22,000.00  $18,217.42  $3,782.58  │
│                           purchase  $60,000.00  $66,051.53 $-6,051.53  │
│                           rental    $10,000.00   $9,481.81    $518.19  │
│                           sets      $15,000.00  $16,422.30 $-1,422.30  │
│                           tape      $20,000.00  $18,256.15  $1,743.85  │
│                                    ----------- ----------- ---------- │
│                                   $207,000.00 $208,429.21 $-1,429.21  │
│                                                                       │
├RKEYS──────────────────────────────────────────────────────────────────┤
│Add_above <F1>    Add_below <F2>     Add_left <F3>     Add_right <F4>   │
│Break     <F5>    CGrow     <F6>     CShrink  <F7>     Define   <F8>    │
│Delete    <F9>    Move    <SHF F0>   Page   <SHF F1>   Pageback <SHF F2>│
│                                                                      R│
└────────────────────────────────────────────────────────────────────────┘
```

Saving the Report Definition and Terminating the REPORT Procedure

Use the RSTORE command (under **File** in the action bar at the top of the REPORT window) to store this report definition as SASUSER.TUTORS.CHAP4.REPT. Then issue the QUIT command to terminate the REPORT procedure. If you want to print the report, follow the directions in Chapter 2, "Creating a Simple Report."

Using the REPORT Language to Produce the Report

The following SAS statements produce the report you developed in this tutorial. For a detailed explanation of these statements, see "Example 3: Grouping and Summarizing Observations" in Chapter 10, "The REPORT Language."

```
proc report data=report.budget;
   title;
   column ('Year-to-Date Financial Status' 'Grouped by Department'
           ' '  dept account budget actual balance);

   define dept    / group format=$10. width=10 'Department';
   define account / group format=$8. width=8 'Account';
   define budget  / sum format=dollar11.2 width=11
                    'Amount/Budgeted';
   define actual  / sum format=dollar11.2 width=11
                    'Amount/Spent';
   define balance / computed format=dollar11.2 width=11
                    'Balance';
```

```
compute balance;
   balance=budget.sum-actual.sum;
endcomp;

break after dept / ol ul skip summarize suppress color=red;

rbreak after / dol dul summarize color=red;

run;
```

Chapter **5** Using Some Advanced Features of the REPORT Procedure

Introduction

In this chapter you create two reports. In the first report, you create a different column for each value of a variable. The second report extends the first one, introducing you to some advanced features of the REPORT procedure. As you create these two reports, you learn how to perform these tasks:

□ move a variable to a new position in the report

□ use a variable so that each value of the variable forms a column in the report (for example, creating one column for each department)

□ place variables below another variable

□ suppress the display of an item in the report

□ add a character variable to the report

□ perform calculations and display the desired results in break lines.

The first report you will create appears in Output 5.1. This report groups the data by DEPT and ACCOUNT as did the report you created in the previous tutorial. However, unlike any report you have created so far, this report uses QTR as an across variable, so that each value of QTR forms a column in the report. In addition, this report uses the variable BUDGET in two distinct ways: once below QTR to show values of BUDGET for each QTR, and once in a column by itself where it represents the sum of BUDGET for both quarters.

Output 5.1
Report with One Column for Each Value of QTR

```
                                         _____QTR_____                          1
                                            1st            2nd
            DEPT        ACCOUNT           BUDGET         BUDGET         BUDGET
            Equipment   lease         $40,000.00     $40,000.00     $80,000.00
                        maint         $10,000.00     $12,000.00     $22,000.00
                        purchase      $40,000.00     $20,000.00     $60,000.00
                        rental         $4,000.00      $6,000.00     $10,000.00
                        sets           $7,500.00      $7,500.00     $15,000.00
                        tape           $8,000.00     $12,000.00     $20,000.00
            Facilities  rent          $24,000.00     $24,000.00     $48,000.00
                        supplies       $2,750.00      $2,750.00      $5,500.00
                        utils          $5,000.00      $3,500.00      $8,500.00
            Other       advert        $30,000.00     $30,000.00     $60,000.00
                        musicfee       $3,000.00      $5,000.00      $8,000.00
                        talent        $13,500.00     $19,500.00     $33,000.00
            Staff       fulltime     $130,000.00    $165,000.00    $295,000.00
                        parttime      $40,000.00     $60,000.00    $100,000.00
            Travel      gas              $800.00      $1,200.00      $2,000.00
                        leases         $3,500.00      $4,500.00      $8,000.00
```

Notice that this report formats the values of QTR, using 1st and 2nd instead of 1 and 2.

Starting the REPORT Procedure

The report you are about to create is not based on any of the reports you previously created. Therefore, you want to start the REPORT procedure without a report definition. To begin with, you want only three variables: QTR, DEPT, and ACCOUNT.

The appropriate statements for starting the procedure follow. The PROC FORMAT step that precedes the PROC REPORT step creates a temporary format that you will use later to display values of 1 as 1st and values of 2 as 2nd. This format makes the reports you create in this tutorial easier to read.

Notice the TITLE statement and the COLUMN statement in the PROC REPORT step. The TITLE statement removes any titles that are in effect. The COLUMN statement selects the three variables to use in the initial display and determines the order in which they appear. Chapter 10, "The REPORT Language," discusses the COLUMN statement in detail.

```
proc format;
    value forqtr 1='1st' 2='2nd';
run;

proc report data=report.budget windows;
    title;
    column qtr dept account;
run;
```

These statements create the report in Display 5.1.

Display 5.1
Report with Three
Variables

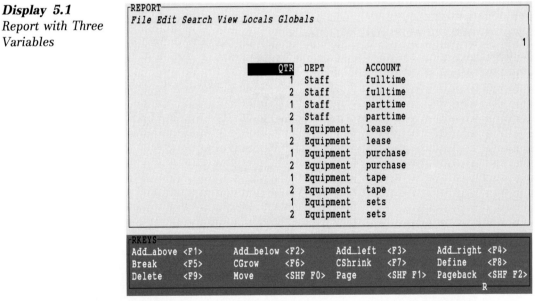

Moving a Variable

Comparing Display 5.1 to Output 5.1 shows that you need to move QTR from the left of DEPT to the right of ACCOUNT. Select **QTR** and issue the MOVE command. (Unless otherwise specified, the commands you use in this chapter are under **Edit** in the action bar at the top of the REPORT window.) When you issue the MOVE command, the REPORT procedure prompts you on how to proceed, as Display 5.2 shows.

Display 5.2
Moving a Variable

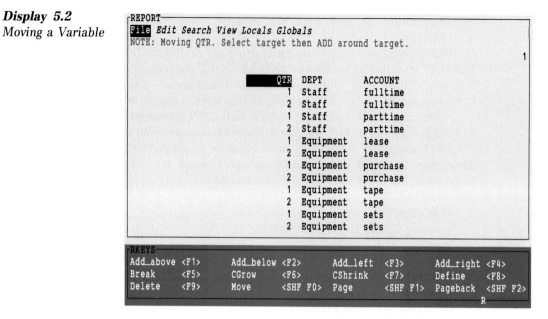

```
┌REPORT──────────────────────────────────────────────────────────────┐
│▓File▓ Edit Search View Locals Globals                               │
│NOTE: Moving QTR. Select target then ADD around target.              │
│                                                                    1│
│                                                                     │
│                        ▓▓▓QTR▓  DEPT        ACCOUNT                  │
│                              1  Staff       fulltime                │
│                              2  Staff       fulltime                │
│                              1  Staff       parttime                │
│                              2  Staff       parttime                │
│                              1  Equipment   lease                   │
│                              2  Equipment   lease                   │
│                              1  Equipment   purchase                │
│                              2  Equipment   purchase                │
│                              1  Equipment   tape                    │
│                              2  Equipment   tape                    │
│                              1  Equipment   sets                    │
│                              2  Equipment   sets                    │
│                                                                     │
└─────────────────────────────────────────────────────────────────────┘
┌RKEYS────────────────────────────────────────────────────────────────┐
│Add_above <F1>      Add_below <F2>      Add_left  <F3>     Add_right <F4>│
│Break     <F5>      CGrow     <F6>      CShrink   <F7>     Define    <F8>│
│Delete    <F9>      Move    <SHF F0>    Page    <SHF F1>   Pageback <SHF F2>│
│                                                                    R│
└─────────────────────────────────────────────────────────────────────┘
```

You want to move QTR to the right of ACCOUNT. Therefore, select **ACCOUNT** as your target and issue the ADD_RIGHT command. The new version of the report, with QTR in its new position, appears in Display 5.3.

Display 5.3
Report with QTR
in Its New Position

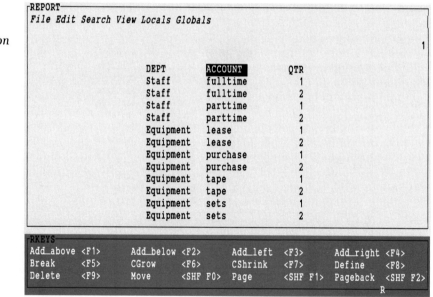

```
┌REPORT──────────────────────────────────────────────────────────────┐
│File Edit Search View Locals Globals                                 │
│                                                                    1│
│                                                                     │
│                    DEPT        ▓ACCOUNT▓     QTR                    │
│                    Staff       fulltime       1                     │
│                    Staff       fulltime       2                     │
│                    Staff       parttime       1                     │
│                    Staff       parttime       2                     │
│                    Equipment   lease          1                     │
│                    Equipment   lease          2                     │
│                    Equipment   purchase       1                     │
│                    Equipment   purchase       2                     │
│                    Equipment   tape           1                     │
│                    Equipment   tape           2                     │
│                    Equipment   sets           1                     │
│                    Equipment   sets           2                     │
│                                                                     │
└─────────────────────────────────────────────────────────────────────┘
┌RKEYS────────────────────────────────────────────────────────────────┐
│Add_above <F1>      Add_below <F2>      Add_left  <F3>     Add_right <F4>│
│Break     <F5>      CGrow     <F6>      CShrink   <F7>     Define    <F8>│
│Delete    <F9>      Move    <SHF F0>    Page    <SHF F1>   Pageback <SHF F2>│
│                                                                    R│
└─────────────────────────────────────────────────────────────────────┘
```

Using the DEFER Mode

As you learned in the preceding tutorial, PROC REPORT cannot create groups if the report contains any display variables that do not have a statistic above or below them or any order variables. It is, therefore, sometimes convenient to turn on DEFER mode when you are creating groups and setting characteristics for several variables at the same time, as you are about to do. DEFER mode suppresses the application of changes to the report (and the accompanying messages) until you are ready to apply them. To turn on DEFER mode, issue the ROPTIONS command (under **Locals** in the action bar at the top of the REPORT window). When the ROPTIONS window appears on your display, select DEFER. When your display looks like Display 5.4, exit the ROPTIONS window.

Display 5.4
Turning On
DEFER Mode

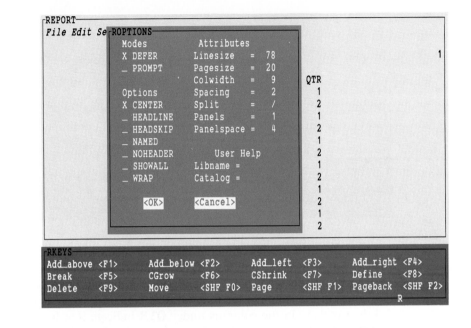

Setting Variable Characteristics

In this section you set the characteristics for the variables in the report. You want DEPT and ACCOUNT once again to be group variables. Use the DEFINE command to set the usage for these variables. Select a variable, issue the DEFINE command, and select **GROUP** in the **Usage** column. The report does not change because you are using DEFER mode.

In this report, you want to create a column for each value of QTR and, eventually, to display values for BUDGET in these columns. To do so, you must first set the usage of QTR to across.

Select **QTR** and issue the DEFINE command. The DEFINITION window for QTR shows that the variable is currently an analysis variable used to calculate the SUM statistic. Set **Usage** to **ACROSS**. Be sure to delete **SUM** from the **Statistic** field because a statistic is only meaningful for an analysis variable. Also, enter `forqtr3.` for the format. When your display looks like Display 5.5, exit the DEFINITION window.

Display 5.5
*Setting the
Characteristics
for QTR*

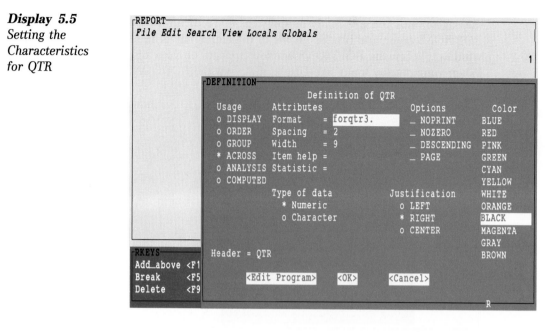

```
┌REPORT─────────────────────────────────────────────────────────┐
│File Edit Search View Locals Globals                            │
│                                                              1 │
│     ┌DEFINITION─────────────────────────────────────────────┐  │
│     │              Definition of QTR                         │  │
│     │  Usage       Attributes           Options      Color   │  │
│     │  o DISPLAY   Format   = forqtr3.  _ NOPRINT    BLUE    │  │
│     │  o ORDER     Spacing  = 2         _ NOZERO     RED     │  │
│     │  o GROUP     Width    = 9         _ DESCENDING PINK    │  │
│     │  * ACROSS    Item help =          _ PAGE       GREEN   │  │
│     │  o ANALYSIS  Statistic =                        CYAN    │  │
│     │  o COMPUTED                                     YELLOW  │  │
│     │              Type of data       Justification  WHITE   │  │
│     │                * Numeric          o LEFT       ORANGE  │  │
│     │                o Character        * RIGHT      BLACK   │  │
│     │                                   o CENTER     MAGENTA │  │
│     │                                                GRAY    │  │
│     │  Header = QTR                                   BROWN   │  │
┌RKEYS──────────┐                                                │  │
│Add_above <F1 │    <Edit Program>    <OK>      <Cancel>        │  │
│Break     <F5 │    └───────────────────────────────────────────┘  │
│Delete    <F9 │                                               R │
└───────────────────────────────────────────────────────────────┘
```

Issue the REFRESH command to see the modified report. As you can see in Display 5.6, the report now uses QTR as an across variable. If, as in this case, an across variable does not have an analysis or a display variable above or below it, the values in the columns for that variable contain a frequency count. In Display 5.6 the numbers in the columns under **QTR** indicate that for each quarter, the report contains one observation for each account in each department.

Display 5.6
*Establishing the
Basic Structure of
the Report*

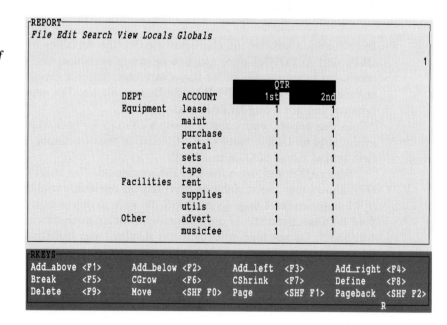

```
┌REPORT─────────────────────────────────────────────────────────┐
│File Edit Search View Locals Globals                            │
│                                                              1 │
│                                          QTR                   │
│                  DEPT        ACCOUNT     1st      2nd           │
│                  Equipment   lease        1        1           │
│                              maint        1        1           │
│                              purchase     1        1           │
│                              rental       1        1           │
│                              sets         1        1           │
│                              tape         1        1           │
│                  Facilities  rent         1        1           │
│                              supplies     1        1           │
│                              utils        1        1           │
│                  Other       advert       1        1           │
│                              musicfee     1        1           │
┌RKEYS──────────────────────────────────────────────────────────┐
│Add_above <F1>    Add_below <F2>    Add_left  <F3>   Add_right <F4>│
│Break     <F5>    CGrow     <F6>    CShrink   <F7>   Define    <F8>│
│Delete    <F9>    Move      <SHF F0> Page     <SHF F1> Pageback <SHF F2>│
└───────────────────────────────────────────────────────────────┘
                                                                R
```

Now that you have created the basic structure for the report and set the characteristics for the variables, issue the ROPTIONS command (under **Locals** in the action bar at the top of the REPORT window) and turn off DEFER mode so that you can immediately see the changes you make to the report.

Adding Variables below a Variable

You are now ready to add BUDGET below QTR. Select **QTR** and issue the ADD_BELOW command. From the ADDING window, select **Dataset variable** and exit the window. From the DATASET VARS window, select **BUDGET**. When your display looks like Display 5.7, exit the window.

Display 5.7
Adding BUDGET below QTR

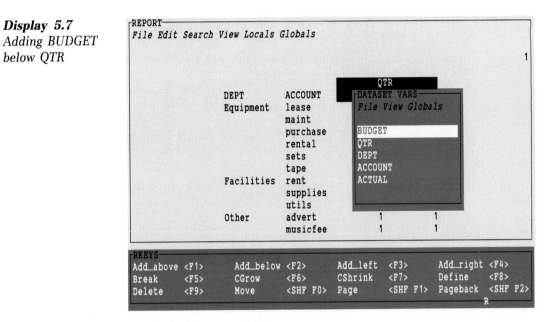

In this report BUDGET is an analysis variable used to calculate the SUM statistic. Therefore, when the REPORT procedure displays the DEFINITION window for BUDGET, accept the default usage and exit the window. The modified report, which includes BUDGET below QTR, appears in Display 5.8.

Display 5.8
Report with BUDGET below QTR

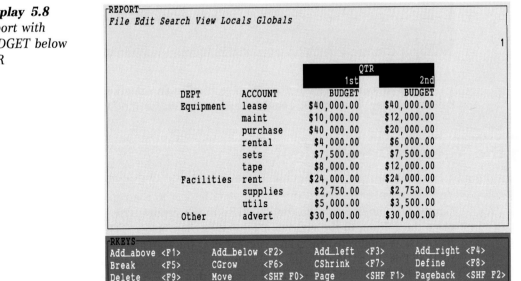

Each row of this report now displays the value of BUDGET for each quarter. For instance, the second row of the report shows that the equipment department budgeted $10,000.00 for maintenance in the first quarter and $12,000.00 for maintenance in the second quarter.

Visually Emphasizing a Column Header

Particularly when a report contains an across variable, you may want to clarify the relationship between the across variable and the variable or variables below it. A simple way to clarify the relationship is to extend the column header of the across variable to cover all the columns below it.

To extend the column header for QTR so that it spans both columns containing the variable BUDGET, type an underscore (_) in the spaces immediately before and immediately after **QTR**. To see the change to the header, you must issue the REFRESH command. The modified report appears in Display 5.9.

Display 5.9
Emphasizing a
Column Header

```
┌REPORT─────────────────────────────────────────────────────────┐
│ File Edit Search View Locals Globals                           │
│                                                              1 │
│                                          ____QTR____            │
│                                          1st        2nd        │
│                  DEPT       ACCOUNT       BUDGET      BUDGET    │
│                  Equipment  lease      $40,000.00  $40,000.00   │
│                             maint      $10,000.00  $12,000.00   │
│                             purchase   $40,000.00  $20,000.00   │
│                             rental      $4,000.00   $6,000.00   │
│                             sets        $7,500.00   $7,500.00   │
│                             tape        $8,000.00  $12,000.00   │
│                  Facilities rent       $24,000.00  $24,000.00   │
│                             supplies    $2,750.00   $2,750.00   │
│                             utils       $5,000.00   $3,500.00   │
│                  Other      advert     $30,000.00  $30,000.00   │
│                                                                │
├RKEYS───────────────────────────────────────────────────────────┤
│ Add_above <F1>      Add_below <F2>     Add_left  <F3>     Add_right <F4> │
│ Break     <F5>      CGrow     <F6>     CShrink   <F7>     Define    <F8> │
│ Delete    <F9>      Move     <SHF F0>  Page     <SHF F1>  Pageback <SHF F2> │
│                                                                R │
└────────────────────────────────────────────────────────────────┘
```

Note: You can use other characters besides an underscore to extend a column header. For details, see the documentation for the DEFINITION window.

Adding the Year-to-Date Budget

You are now ready to add the year-to-date budget to the report. To do so, you simply add another column for BUDGET to the report. This time, however, you place the column to the right of QTR rather than below QTR. In this position, BUDGET is unaffected by QTR. Therefore, if the statistic you associate with BUDGET is the SUM statistic, the values displayed in this column represent the sum of BUDGET for all rows of the report that have a unique combination of values of DEPT and ACCOUNT.

To add BUDGET in this position, select **QTR** and issue the ADD_RIGHT command. From the ADDING window, select **Dataset variable** and exit the window. From the DATASET VAR window, select **BUDGET** and exit the window. When the DEFINITION window for BUDGET appears, simply accept the default usage by exiting the window.

The modified report appears in Display 5.10.

Display 5.10
Report with Two
Uses of BUDGET

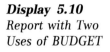

```
┌REPORT─────────────────────────────────────────────────────────────┐
│ File Edit Search View Locals Globals                               │
│                                                                  1 │
│                                                                    │
│                                        ┌────QTR────┐               │
│                                        │1st    2nd │               │
│              DEPT        ACCOUNT        BUDGET      BUDGET    BUDGET│
│              Equipment   lease       $40,000.00  $40,000.00 $80,000.00│
│                          maint       $10,000.00  $12,000.00 $22,000.00│
│                          purchase    $40,000.00  $20,000.00 $60,000.00│
│                          rental       $4,000.00   $6,000.00 $10,000.00│
│                          sets         $7,500.00   $7,500.00 $15,000.00│
│                          tape         $8,000.00  $12,000.00 $20,000.00│
│              Facilities  rent        $24,000.00  $24,000.00 $48,000.00│
│                          supplies     $2,750.00   $2,750.00  $5,500.00│
│                          utils        $5,000.00   $3,500.00  $8,500.00│
│              Other       advert      $30,000.00  $30,000.00 $60,000.00│
└────────────────────────────────────────────────────────────────────┘
┌RKEYS───────────────────────────────────────────────────────────────┐
│ Add_above <F1>    Add_below <F2>    Add_left <F3>    Add_right <F4> │
│ Break     <F5>    CGrow     <F6>    CShrink  <F7>    Define   <F8>  │
│ Delete    <F9>    Move    <SHF F0>  Page   <SHF F1>  Pageback <SHF F2>│
│                                                                   R │
└────────────────────────────────────────────────────────────────────┘
```

Notice that in each row of this report, the rightmost value of BUDGET consolidates values of BUDGET for each group of observations in the data set that have a unique combination of values for DEPT and ACCOUNT (that is, for two quarters). The values of BUDGET that are below QTR represent the value for a single quarter only. For instance, the first detail line of the report shows that the equipment department budgeted $40,000 for leasing in the first quarter, $40,000 for leasing in the second quarter, and a total of $80,000 for leasing for both quarters.

You have completed the first report in this tutorial. You can stop here if you want to. Use the RSTORE command to store this report definition in SASUSER.TUTORS.CHAP5A.REPT. The remainder of this tutorial elaborates on the report you just completed to produce the more sophisticated report in Output 5.2.

Output 5.2
Elaborating on the
Previous Report

```
┌─────────────────────────────────────────────────────────────────────┐
│                          ────────Quarter────────                     │
│                              1st         2nd       Year-to-Date       │
│          Department Account  Balance     Balance   Balance            │
│          Equipment  lease       $0.00       $0.00      $0.00          │
│                     maint    $2,457.87   $1,324.71  $3,782.58         │
│                     purchase $-8,282.38  $2,230.85 $-6,051.53*        │
│                     rental       $1.13     $517.06    $518.19         │
│                     sets      $-842.68    $-579.62 $-1,422.30*        │
│                     tape     $1,170.58     $573.27  $1,743.85         │
│          ──────────          ──────────  ────────── ─────────────     │
│          Equipment          $-5,495.48   $4,066.27 $-1,429.21*        │
│                                                                       │
│          27% of the year-to-date budget is allocated to this department.│
└─────────────────────────────────────────────────────────────────────┘
```

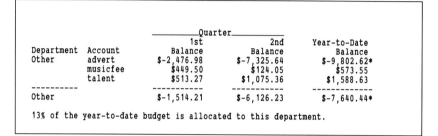

```
                                _____Quarter_____
                                    1st           2nd       Year-to-Date
            Department  Account   Balance       Balance        Balance
            Facilities  rent        $0.00         $0.00          $0.00
                        supplies  $533.45         $7.52        $540.97
                        utils     $776.71        $55.19        $831.90
            ----------            -----------   -----------   ------------
            Facilities           $1,310.16       $62.71      $1,372.87

              8% of the year-to-date budget is allocated to this department.
```

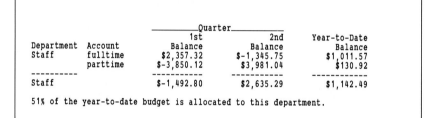

```
                                _____Quarter_____
                                    1st           2nd       Year-to-Date
            Department  Account   Balance       Balance        Balance
            Other       advert  $-2,476.98    $-7,325.64    $-9,802.62*
                        musicfee  $449.50       $124.05        $573.55
                        talent    $513.27     $1,075.36      $1,588.63
            ----------            -----------   -----------   ------------
            Other                $-1,514.21    $-6,126.23    $-7,640.44*

              13% of the year-to-date budget is allocated to this department.
```

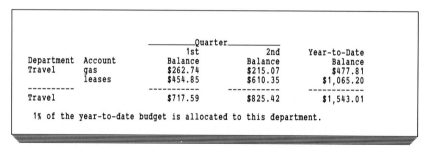

```
                                _____Quarter_____
                                    1st           2nd       Year-to-Date
            Department  Account   Balance       Balance        Balance
            Staff       fulltime $2,357.32    $-1,345.75     $1,011.57
                        parttime $-3,850.12    $3,981.04       $130.92
            ----------            -----------   -----------   ------------
            Staff                $-1,492.80    $2,635.29     $1,142.49

              51% of the year-to-date budget is allocated to this department.
```

```
                                _____Quarter_____
                                    1st           2nd       Year-to-Date
            Department  Account   Balance       Balance        Balance
            Travel      gas       $262.74       $215.07        $477.81
                        leases    $454.85       $610.35      $1,065.20
            ----------            -----------   -----------   ------------
            Travel                $717.59       $825.42      $1,543.01

              1% of the year-to-date budget is allocated to this department.
```

The report in Output 5.2 uses some advanced features of PROC REPORT to produce a simple-looking but sophisticated report. This report is more complex than the reports you have produced so far. Therefore, you are apt to need more time to complete it. Remember that at any time, you can use the RSTORE command to store your current report definition and subsequently resume the tutorial at that point.

Understanding the Structure of the Sophisticated Report

The report in Output 5.2 is more sophisticated than it appears to be. This section describes the structure of the report. It is important to understand the structure before you begin building the report.

Report Variables

Like the report you just completed, this report uses DEPT and ACCOUNT as group variables and QTR as an across variable. Below QTR is a computed variable, BALANCE, which shows the quarterly balance for each account. To the right of QTR is another computed variable, YRTODATE (with a column header of 'Year-to-Date/Balance'). This variable shows the balance for each account over both quarters combined. The report also contains a character variable to the right of YRTODATE. The column header for this variable is blank. The value of this variable is an asterisk (*) if YRTODATE is less than 0 and a blank if YRTODATE is greater than 0. Thus, the variable serves to flag accounts that have spent more than their budgets.

The calculations of the computed variables BALANCE and YRTODATE are based on variables whose display is suppressed in the final report. Output 5.3 shows the first page of the report with all the variables displayed. (Note that the boldface numbers are not part of the report but simply identify column numbers for later discussion.)

Output 5.3   *Output with No Columns Suppressed*

```
                                              Quarter                                                                        11
                 1           2             1st                        2nd                                                Year-to-Date
            Department  Account    3 BUDGET   4 ACTUAL   5 Balance   6 BUDGET   7 ACTUAL    8 Balance  9 BUDGET   10 ACTUAL    Balance
            Equipment   lease      $40,000.00 $40,000.00      $0.00 $40,000.00 $40,000.00       $0.00 $80,000.00 $80,000.00       $0.00
                        maint      $10,000.00  $7,542.13  $2,457.87 $12,000.00 $10,675.29   $1,324.71 $22,000.00 $18,217.42   $3,782.58
                        purchase   $40,000.00 $48,282.38 $-8,282.38 $20,000.00 $17,769.15   $2,230.85 $60,000.00 $66,051.53  $-6,051.53*
                        rental      $4,000.00  $3,998.87      $1.13  $6,000.00  $5,482.94     $517.06 $10,000.00  $9,481.81     $518.19
                        sets        $7,500.00  $8,342.68   $-842.68  $7,500.00  $8,079.62    $-579.62 $15,000.00 $16,422.30  $-1,422.30*
                        tape        $8,000.00  $6,829.42  $1,170.58 $12,000.00 $11,426.73     $573.27 $20,000.00 $18,256.15   $1,743.85
            ----------  ----------  ---------- ---------- ---------- ---------- ----------  ---------- ---------- ---------- ----------
            Equipment            $109,500.00 $114,995.48 $-5,495.48 $97,500.00 $93,433.73   $4,066.27 $207,000.00 $208,429.21 $-1,429.21*

            27% of the year-to-date budget is allocated to this department.
```

Throughout this tutorial, it is important to realize that just as the previous report uses BUDGET in two different ways, this report uses both BUDGET and ACTUAL in two ways. When the variables are below QTR, as they are in columns 3, 4, 6, and 7, the values in each column are the values for one quarter. When the variables are to the right of QTR (as in columns 9 and 10), the values in the columns are the sums of the values for both quarters. The REPORT procedure uses the values for BUDGET and ACTUAL that are below QTR to calculate each quarterly balance. Similarly, it uses the values for BUDGET and ACTUAL that are to the right of QTR to calculate the year-to-date balance.

For instance, the last detail row of the report shows that the tape account in the equipment department

□ had a budget of $8,000 during the first quarter and spent $6,829.42 during that quarter, leaving a balance for the quarter of $1,170.58.

□ had a budget of $12,000 during the second quarter and spent $11,426.73 during that quarter, leaving a balance for the quarter of $573.27.

□ had a budget of $20,000.00 for both quarters (or for the year to date) and spent $18,256.15 over both quarters, leaving a year-to-date balance of $1,743.85.

Customizing Break Lines

One important difference between the report in Output 5.2 and the other reports you have seen is the line of text after the rows for each department. This line reports the percentage of the year-to-date budget allocated to the department. In order to calculate the percentage you must determine the total budget for the company and the budget for each department.

You create a customized break line to display this information by using DATA step statements in the COMPUTE window associated with the break in the report.

Controlling the Layout of a Report

As Output 5.3 illustrates, the intermediate reports you use to create the final report in this tutorial contain many variables. It is easier to work with these reports if the variables all fit on one line. It is also easier if you don't center the report because then, regardless of the number of variables you are working with at the moment, you can always see some of them on your display.

By default, PROC REPORT determines what line size to use and whether or not to center the report from the LS= and CENTER options in the PROC REPORT statement. If you don't use these options, PROC REPORT uses the settings stored in a report loaded with the REPORT= option. If you aren't using a report definition, the values of the corresponding system options are used. You can change these values for the duration of a PROC REPORT session with two options in the ROPTIONS window. To do so, select **Locals** from the action bar at the top of the REPORT window. From the pull-down menu that appears, select **ROPTIONS**. In the ROPTIONS window, cancel the **CENTER** option and enter a value of **255** in the **Linesize** field. When your display looks like Display 5.11, exit the ROPTIONS window. (Remember that your **Linesize** and **Pagesize** values may differ from those in Display 5.11.)

Display 5.11
Controlling
Centering and Line
Size in a REPORT
Session

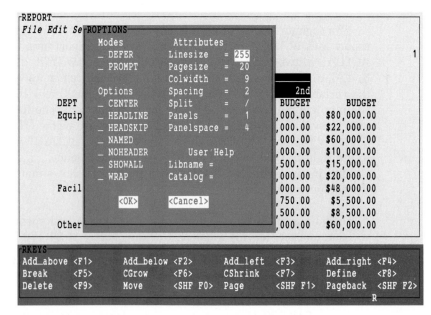

PROC REPORT repositions the report on your display, as Display 5.12 shows.

Display 5.12
Adjusting the
Layout of the
Report

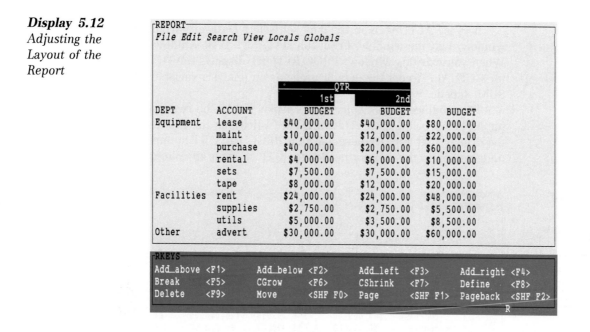

Adding ACTUAL to the Report

As usual, the balances in the report you are creating are based on BUDGET and ACTUAL. The report shows a balance for each quarter and a year-to-date balance. Therefore, you must add ACTUAL to the report in two places: to the right of BUDGET when it is below QTR, and to the right of BUDGET when it is to the right of QTR.

First, select either occurrence of **BUDGET** below **QTR**. Regardless of which of these two columns you select, PROC REPORT highlights them both because they represent one item in the report. (Remember that you added both columns at once when you added BUDGET below QTR earlier in this tutorial.) When your display looks like Display 5.13, issue the ADD_RIGHT command.

Display 5.13
Selecting BUDGET
below QTR

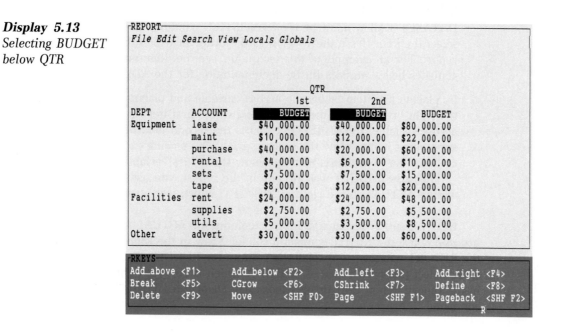

Because ACTUAL is in the data set, select **Dataset variable** from the ADDING window. Exit the window. From the DATASET VAR window, select **ACTUAL**. When you exit this window, PROC REPORT displays the DEFINITION window for ACTUAL. Accept the default usage as an analysis variable used to calculate the SUM statistic.

The modified report appears in Display 5.14. The report contains a column for ACTUAL below each value of QTR. Notice that the report is now wider than the display that appears here and that PROC REPORT automatically extends the underscores in the column header for QTR to span all columns below quarter.

Display 5.14
Report with
ACTUAL below
QTR

```
┌REPORT──────────────────────────────────────────────────────────────────────
│ File Edit Search View Locals Globals
│
│
│                             ───────────────────QTR───────────────────
│                                    1st                    2nd
│                            ┌──────────┐           ┌──────────┐
│         DEPT      ACCOUNT  │  BUDGET  │   ACTUAL  │  BUDGET  │   ACTUAL
│         Equipment lease     $40,000.00  $40,000.00  $40,000.00  $40,000.00
│                   maint     $10,000.00   $7,542.13  $12,000.00  $10,675.29
│                   purchase  $40,000.00  $48,282.38  $20,000.00  $17,769.15
│                   rental     $4,000.00   $3,998.87   $6,000.00   $5,482.94
│                   sets       $7,500.00   $8,342.68   $7,500.00   $8,079.62
│                   tape       $8,000.00   $6,829.42  $12,000.00  $11,426.73
│         Facilities rent     $24,000.00  $24,000.00  $24,000.00  $24,000.00
│                   supplies   $2,750.00   $2,216.55   $2,750.00   $2,742.48
│                   utils      $5,000.00   $4,223.29   $3,500.00   $3,444.81
│         Other     advert    $30,000.00  $32,476.98  $30,000.00  $37,325.64
└──────────────────────────────────────────────────────────────────────────────
┌RKEYS─────────────────────────────────────────────────────────────────────────
│ Add_above <F1>      Add_below <F2>      Add_left  <F3>      Add_right <F4>
│ Break     <F5>      CGrow     <F6>      CShrink   <F7>      Define    <F8>
│ Delete    <F9>      Move      <SHF F0>  Page      <SHF F1>  Pageback  <SHF F2>
│                                                                          R
```

Note: You can display multiple items in adjacent columns below an across variable in one of two ways:

□ by issuing the ADD_BELOW command and adding all items at once from the DATASET VAR window or the STATISTICS window. The order in which you select the items from the window is the order in which they appear in the report. For an example of this technique, see the illustration of adding statistics below an item in the documentation for the ADD_BELOW command.

□ by placing one item below the across variable and placing the second item to the right or left of the first item. Because the first item is already below the across variable, PROC REPORT places the second item below it also. Place subsequent items below the across variable by placing each item to the right or left of an item that is already below the across variable.

Because BUDGET was already in the report, you use this method here, adding ACTUAL to the right of BUDGET, which was already below QTR.

Now you are ready to add another column for ACTUAL to the report. If necessary, use the RIGHT command (under **View** in the action bar at the top of the REPORT window) to display the column for BUDGET that is to the right of QTR. Select this occurrence of BUDGET, which is currently the rightmost column in the report. When your display looks like Display 5.15, issue the ADD_RIGHT command.

Display 5.15
Selecting the
Occurrence of
BUDGET That Is
to the Right of
QTR

```
┌REPORT──────────────────────────────────────────────────────────┐
│ File Edit Search View Locals Globals                            │
│                                                                 │
│                                                                 │
│                                                                 │
│         ─────QTR──────────────────────────────                  │
│                       2nd                                       │
│          ACTUAL       BUDGET        ACTUAL         BUDGET        │
│       40,000.00   $40,000.00    $40,000.00    $80,000.00         │
│       $7,542.13   $12,000.00    $10,675.29    $22,000.00         │
│      48,282.38    $20,000.00    $17,769.15    $60,000.00         │
│       $3,998.87    $6,000.00     $5,482.94    $10,000.00         │
│       $8,342.68    $7,500.00     $8,079.62    $15,000.00         │
│       $6,829.42   $12,000.00    $11,426.73    $20,000.00         │
│      24,000.00    $24,000.00    $24,000.00    $48,000.00         │
│       $2,216.55    $2,750.00     $2,742.48     $5,500.00         │
│       $4,223.29    $3,500.00     $3,444.81     $8,500.00         │
│      32,476.98    $30,000.00    $37,325.64    $60,000.00         │
│                                                                 │
├RKEYS────────────────────────────────────────────────────────────┤
│ Add_above <F1>    Add_below <F2>    Add_left  <F3>   Add_right <F4> │
│ Break     <F5>    CGrow     <F6>    CShrink   <F7>   Define    <F8> │
│ Delete    <F9>    Move   <SHF F0>   Page   <SHF F1>  Pageback <SHF F2> │
│                                                               R │
└─────────────────────────────────────────────────────────────────┘
```

Once again select **Dataset variable** from the ADDING window and select **ACTUAL** from the DATASET VAR window. In the DEFINITION window for ACTUAL, accept the default usage.

The modified report appears in Display 5.16. All the data set variables you need are now in the report, although you may need to use the RIGHT command to see them all.

Display 5.16
Report with All
Data Set Variables
in Place

```
┌REPORT──────────────────────────────────────────────────────────┐
│ File Edit Search View Locals Globals                            │
│                                                                 │
│                                                                 │
│                                                                 │
│         ─────QTR──────────────────────────────                  │
│                       2nd                                       │
│          ACTUAL       BUDGET        ACTUAL         BUDGET        ACTUAL │
│       40,000.00   $40,000.00    $40,000.00    $80,000.00    $80,000.00 │
│       $7,542.13   $12,000.00    $10,675.29    $22,000.00    $18,217.42 │
│      48,282.38    $20,000.00    $17,769.15    $60,000.00    $66,051.53 │
│       $3,998.87    $6,000.00     $5,482.94    $10,000.00     $9,481.81 │
│       $8,342.68    $7,500.00     $8,079.62    $15,000.00    $16,422.30 │
│       $6,829.42   $12,000.00    $11,426.73    $20,000.00    $18,256.15 │
│      24,000.00    $24,000.00    $24,000.00    $48,000.00    $48,000.00 │
│       $2,216.55    $2,750.00     $2,742.48     $5,500.00     $4,959.03 │
│       $4,223.29    $3,500.00     $3,444.81     $8,500.00     $7,668.10 │
│      32,476.98    $30,000.00    $37,325.64    $60,000.00    $69,802.62 │
│                                                                 │
├RKEYS────────────────────────────────────────────────────────────┤
│ Add_above <F1>    Add_below <F2>    Add_left  <F3>   Add_right <F4> │
│ Break     <F5>    CGrow     <F6>    CShrink   <F7>   Define    <F8> │
│ Delete    <F9>    Move   <SHF F0>   Page   <SHF F1>  Pageback <SHF F2> │
│                                                               R │
└─────────────────────────────────────────────────────────────────┘
```

Adding the Year-to-Date Balance

You are now ready to add the year-to-date balance to the report. To do so, you add a computed variable to the right of ACTUAL (the occurrence of ACTUAL that you just added), defining the variable as the difference between the values for BUDGET and ACTUAL in that line of the report. You did something similar in the previous tutorial.

Defining the Variable

Issue the RIGHT command if necessary, and select the occurrence of **ACTUAL** that is in the rightmost column of the report. Next, issue the ADD_RIGHT command. From the ADDING window select **Computed variable** and exit the window. Enter the name `yrtodate` at the prompt in the COMPUTED VAR window. Because `yrtodate` is a numeric variable, you can ignore the **Character data** check box and the **Length** prompt. When your display looks like Display 5.17, select the **Edit Program** push button.

Display 5.17
Adding a
Computed Variable

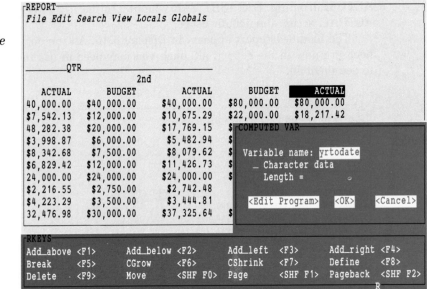

After you select the **Edit Program** push button, PROC REPORT displays the COMPUTE window for the new variable, YRTODATE. Because BUDGET and ACTUAL are analysis variables used to calculate the SUM statistic, you must use compound names in the definition, as explained in "Computing a Variable Based on Statistics" in Chapter 4, "Grouping and Summarizing Data." The assignment statement to use appears in Display 5.18. When your display looks like Display 5.18, exit the COMPUTE window and the COMPUTED VAR window.

Display 5.18
*Using Compound
Names to Define a
Computed Variable*

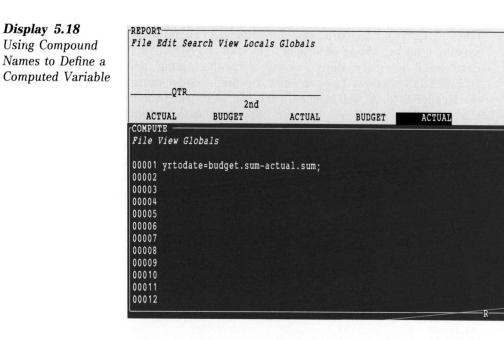

The modified report includes the new variable **YRTODATE**. Display 5.19 shows the rightmost part of the report.

Display 5.19
*Report with
Year-to-Date
Balance*

```
┌REPORT─────────────────────────────────────────────────────┐
│ ███ Edit Search View Locals Globals                        │
│                                                            │
│                                                            │
│        ____QTR_____             │
│                      2nd                                   │
│      ACTUAL         BUDGET          ACTUAL        BUDGET          ACTUAL    YRTODATE │
│  40,000.00      $40,000.00      $40,000.00    $80,000.00      $80,000.00           0 │
│  $7,542.13      $12,000.00      $10,675.29    $22,000.00      $18,217.42        3783 │
│  48,282.38      $20,000.00      $17,769.15    $60,000.00      $66,051.53       -6052 │
│  $3,998.87       $6,000.00       $5,482.94    $10,000.00       $9,481.81         518 │
│  $8,342.68       $7,500.00       $8,079.62    $15,000.00      $16,422.30       -1422 │
│  $6,829.42      $12,000.00      $11,426.73    $20,000.00      $18,256.15        1744 │
│  24,000.00      $24,000.00      $24,000.00    $48,000.00      $48,000.00           0 │
│  $2,216.55       $2,750.00       $2,742.48     $5,500.00       $4,959.03         541 │
│  $4,223.29       $3,500.00       $3,444.81     $8,500.00       $7,668.10         832 │
│  32,476.98      $30,000.00      $37,325.64    $60,000.00      $69,802.62       -9803 │
│                                                            │
├RKEYS───────────────────────────────────────────────────────┤
│ Add_above  <F1>      Add_below  <F2>      Add_left  <F3>      Add_right  <F4> │
│ Break      <F5>      CGrow      <F6>      CShrink   <F7>      Define     <F8> │
│ Delete     <F9>      Move       <SHF F0>  Page      <SHF F1>  Pageback   <SHF F2> │
└──────────────────────────────────────────────────────────R─┘
```

Setting the Characteristics of YRTODATE

Recall from previous tutorials that when you add a computed variable, the REPORT procedure uses default values for various attributes, including the format and width. In this case, you want to set the format of YRTODATE to match the format of the other numeric variables in the report. You also want to use a column that is wider than the default to accommodate not only the format (which takes a maximum of 11 spaces) but also the column header you will use (which takes 12 spaces). Previously, you have changed column headers by typing over the existing headers, adding additional lines with the ADD_ABOVE or ADD_BELOW command if necessary. Here, you will define the column header in the DEFINITION window. You can enter a column header of up to 40 characters in the **Header** field. As you learned when you were using the PROMPT facility, the slash (/) is the default split character, which splits the column header over multiple lines.

Select **YRTODATE** and issue the DEFINE command. In the DEFINITION window for YRTODATE, set the format to `dollar11.2` and the width to `12`. Finally, enter the text `Year-to-Date/Balance` in the **Header** field.

When your display looks like Display 5.20, exit the window.

Display 5.20
Setting
Characteristics for
YRTODATE

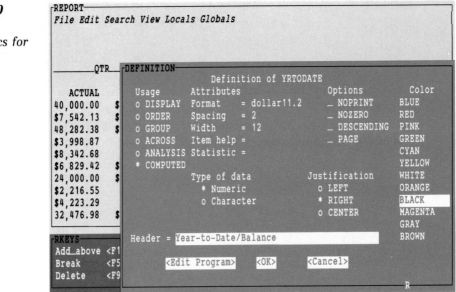

Display 5.21 shows the rightmost part of the report, which includes YRTODATE with its new format, width, and column header.

Display 5.21
Report with New Characteristics for YRTODATE

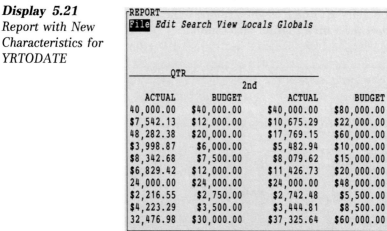

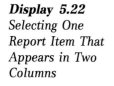

Adding Quarterly Balances

To calculate a balance for each quarter, you need to add a variable to the right of the occurrence of ACTUAL that is below QTR and to define that variable as the difference between the values of BUDGET and ACTUAL that are below QTR. To add this variable, select either of the columns below QTR that has the column header **ACTUAL**. As Display 5.22 shows, when you select either of these columns, the REPORT procedure highlights them both because you are selecting a single report item.

Display 5.22
Selecting One Report Item That Appears in Two Columns

```
┌REPORT─────────────────────────────────────────────────────────────────┐
│ File Edit Search View Locals Globals                                   │
│                                                                        │
│                                                                        │
│          ───QTR───                                                     │
│                       2nd                            Year-to-Date      │
│        ▐ACTUAL▌       BUDGET        ▐ACTUAL▌    BUDGET      ACTUAL      Balance │
│       40,000.00    $40,000.00    $40,000.00  $80,000.00  $80,000.00      $0.00  │
│        $7,542.13    $12,000.00    $10,675.29  $22,000.00  $18,217.42  $3,782.58 │
│       48,282.38    $20,000.00    $17,769.15  $60,000.00  $66,051.53 $-6,051.53  │
│        $3,998.87     $6,000.00     $5,482.94  $10,000.00   $9,481.81    $518.19  │
│        $8,342.68     $7,500.00     $8,079.62  $15,000.00  $16,422.30 $-1,422.30 │
│        $6,829.42    $12,000.00    $11,426.73  $20,000.00  $18,256.15  $1,743.85 │
│       24,000.00    $24,000.00    $24,000.00  $48,000.00  $48,000.00      $0.00  │
│        $2,216.55     $2,750.00     $2,742.48   $5,500.00   $4,959.03    $540.97  │
│        $4,223.29     $3,500.00     $3,444.81   $8,500.00   $7,668.10    $831.90  │
│       32,476.98    $30,000.00    $37,325.64  $60,000.00  $69,802.62 $-9,802.62  │
│                                                                        │
├RKEYS───────────────────────────────────────────────────────────────────┤
│ Add_above <F1>      Add_below <F2>      Add_left  <F3>    Add_right <F4> │
│ Break     <F5>      CGrow     <F6>      CShrink   <F7>    Define    <F8> │
│ Delete    <F9>      Move      <SHF F0>  Page    <SHF F1>  Pageback <SHF F2> │
│                                                                      R  │
└────────────────────────────────────────────────────────────────────────┘
```

Issue the ADD_RIGHT command. From the ADDING window, select
Computed variable and exit the window. When the COMPUTED VAR window

appears, enter the name `qtrbal` at the prompt. When your display looks like Display 5.23, select the **Edit Program** push button.

Display 5.23
Adding Quarterly
Balances

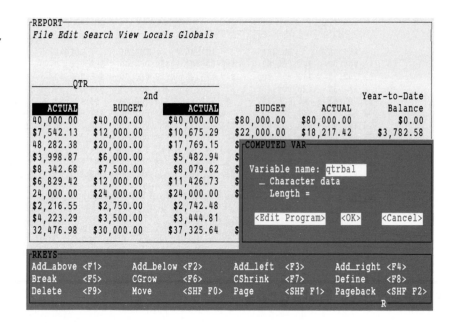

Referring to Report Values by Column Number

In previous reports you used variable names and compound names to refer to values used in computing a new variable. However, when multiple items share a column, as they do in this report (QTR shares columns with BUDGET and ACTUAL), you cannot use either of these methods to refer to a value. Instead, you must use names of the form _Cn_, where *n* is the number of the column containing the item to use.

When you refer to columns by number, you cannot use a single statement to define a computed variable that appears below an across variable, because values for the variable appear in more than one column. For example, in this report a value for QTRBAL appears under each quarter, to the right of ACTUAL for that quarter. To calculate the balance for the first quarter, which will appear in column 5 of the report, subtract the value of ACTUAL in column 4 from the value of BUDGET in column 3 (refer to Output 5.3 for a picture of the report with numbered columns):

    _c5_=_c3_-_c4_;

To calculate a balance for the second quarter, use the following statement:

    _c8_=_c6_-_c7_;

Note: Although the values of BUDGET and ACTUAL for the second quarter are currently in columns 5 and 6 of the report, after the addition of the first quarterly balance, they will be in columns 6 and 7.

These, then, are the statements you use to define the quarterly balances in the COMPUTE window. When your display looks like Display 5.24, exit the COMPUTE window and the COMPUTED VAR window.

Display 5.24
Using Column Numbers to Refer to Report Values

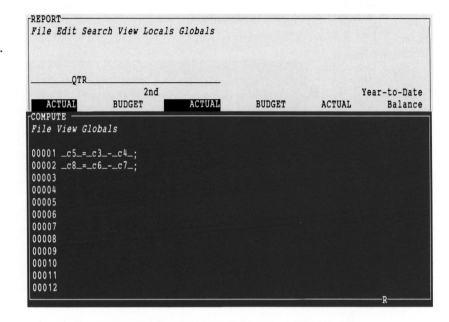

Display 5.25 shows the report with both quarterly balances.

Display 5.25
Report with Quarterly Balances

Setting Characteristics for QTRBAL

Next, you need to set the characteristics for QTRBAL to those used in the report. Select either column containing QTRBAL and issue the DEFINE command. In the DEFINITION window for QTRBAL, set the format to DOLLAR11.2, the width to

11, and the column header to `Balance`. When your display looks like Display 5.26, exit the window.

Display 5.26
Setting
Characteristics for
QTRBAL

```
┌REPORT──────────────────────────────────────────────────────┐
│File Edit Search View Locals Globals                        │
│                                                            │
│                                                            │
│         ┌DEFINITION──────────────────────────────────────┐ │
│ 1st     │             Definition of QTRBAL               │ │
│ ACTUAL  │ Usage     Attributes          Options    Color │ │
│,000.00  │ o DISPLAY Format   = dollar11.2 _ NOPRINT  BLUE │ │
│,542.13  │ o ORDER   Spacing  = 2          _ NOZERO   RED  │ │
│,282.38  │ o GROUP   Width    = 11         _ DESCENDING PINK│ │
│,998.87  │ o ACROSS  Item help =           _ PAGE     GREEN │ │
│,342.68  │ o ANALYSIS Statistic =                     CYAN │ │
│,829.42  │ * COMPUTED                                 YELLOW│ │
│,000.00  │              Type of data    Justification WHITE │ │
│,216.55  │                * Numeric     o LEFT       ORANGE │ │
│,223.29  │                o Character   * RIGHT      BLACK  │ │
│,476.98  │                              o CENTER     MAGENTA│ │
│         │                                           GRAY  │ │
│┌RKEYS── │ Header = Balance                          BROWN │ │
││Add_above <F1│                                            │ │
││Break    <F5│   <Edit Program>   <OK>    <Cancel>         │ │
││Delete   <F9│                                          R  │ │
│         └────────────────────────────────────────────────┘ │
└────────────────────────────────────────────────────────────┘
```

The latest version of the report, which appears in Display 5.27, contains both quarterly balances, now correctly formatted.

Display 5.27
Report with New
Characteristics for
QTRBAL

```
┌REPORT──────────────────────────────────────────────────────┐
│█File█ Edit Search View Locals Globals                      │
│                                                            │
│                                                            │
│                    ─────QTR────────────────────────────    │
│ 1st                          2nd                           │
│ ACTUAL     Balance     BUDGET      ACTUAL     Balance    BUDGET│
│,000.00       $0.00  $40,000.00  $40,000.00      $0.00  $80,000.00│
│,542.13   $2,457.87  $12,000.00  $10,675.29  $1,324.71  $22,000.00│
│,282.38  $-8,282.38  $20,000.00  $17,769.15  $2,230.85  $60,000.00│
│,998.87       $1.13   $6,000.00   $5,482.94    $517.06  $10,000.00│
│,342.68    $-842.68   $7,500.00   $8,079.62   $-579.62  $15,000.00│
│,829.42   $1,170.58  $12,000.00  $11,426.73    $573.27  $20,000.00│
│,000.00       $0.00  $24,000.00  $24,000.00      $0.00  $48,000.00│
│,216.55     $533.45   $2,750.00   $2,742.48      $7.52   $5,500.00│
│,223.29     $776.71   $3,500.00   $3,444.81     $55.19   $8,500.00│
│,476.98  $-2,476.98  $30,000.00  $37,325.64  $-7,325.64 $60,000.00│
│                                                            │
│┌RKEYS───────────────────────────────────────────────────┐ │
││Add_above <F1>    Add_below <F2>   Add_left  <F3>   Add_right <F4>│
││Break     <F5>    CGrow     <F6>   CShrink   <F7>   Define    <F8>│
││Delete    <F9>    Move   <SHF F0>  Page   <SHF F1>  Pageback <SHF F2>│
│└────────────────────────────────────────────────────── R ┘ │
└────────────────────────────────────────────────────────────┘
```

Suppressing the Display of a Variable

The report you are now creating does not show values for the quarterly values of BUDGET and ACTUAL or for the year-to-date values for the same variables. You can't delete them from the report because you need the values to calculate

QTRBAL and YRTODATE. However, you can easily suppress their display. To do so, select **BUDGET** below QTR and issue the DEFINE command. When the REPORT procedure displays the DEFINITION window for BUDGET, select the **NOPRINT** option. When your display looks like Display 5.28, exit the window.

Display 5.28
Suppressing the Display of a Variable

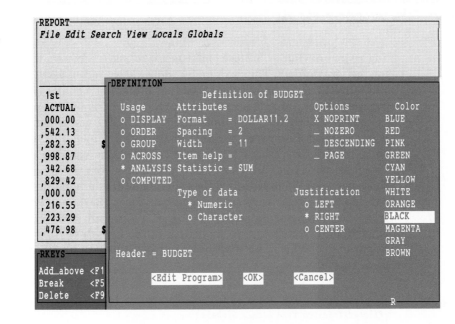

In the same way, suppress the display of ACTUAL under QTR and of the columns for BUDGET and ACTUAL that are to the right of QTR. When you have suppressed the display of these variables, your report looks like Display 5.29.

Display 5.29
Report with All Occurrences of BUDGET and ACTUAL Suppressed

```
┌REPORT─────────────────────────────────────────────────────────────┐
│File Edit Search View Locals Globals                                │
│                                                                    │
│                                                                    │
│                          ─────────QTR─────────                     │
│                               1st        2nd      Year-to-Date     │
│ DEPT        ACCOUNT        Balance    Balance         Balance      │
│ Equipment   lease           $0.00      $0.00           $0.00       │
│             maint       $2,457.87  $1,324.71       $3,782.58       │
│             purchase    $-8,282.38  $2,230.85      $-6,051.53       │
│             rental          $1.13    $517.06         $518.19       │
│             sets         $-842.68   $-579.62      $-1,422.30       │
│             tape        $1,170.58    $573.27       $1,743.85       │
│ Facilities  rent            $0.00      $0.00           $0.00       │
│             supplies     $533.45       $7.52         $540.97       │
│             utils        $776.71      $55.19         $831.90       │
│ Other       advert     $-2,476.98  $-7,325.64      $-9,802.62       │
│                                                                    │
├RKEYS───────────────────────────────────────────────────────────────
│Add_above <F1>     Add_below <F2>     Add_left  <F3>    Add_right <F4>
│Break     <F5>     CGrow     <F6>     CShrink   <F7>    Define    <F8>
│Delete    <F9>     Move   <SHF F0>    Page   <SHF F1>   Pageback <SHF F2>
│                                                                  R
```

Note: Much of your report is now complete. If you want to take a break, remember that you can store the current report definition and later resume the tutorial, starting with that definition. If you do so, be sure to re-create the format FORQTR. before trying to use the report definition, or store the format in a permanent library. For information on storing formats in a permanent library, see Chapter 18, "The FORMAT Procedure," in the *SAS Procedures Guide, Version 6, Third Edition.*

Adding a Character Variable

The report in Output 5.2 uses an asterisk to mark accounts that are over budget for the year to date. This is done by adding a character variable to the right of YRTODATE. The value of the variable is a blank if YRTODATE is greater than or equal to 0. The value is an asterisk if YRTODATE is less than 0.

Select **YRTODATE** and issue the ADD_RIGHT command. From the ADDING window, select **Computed variable** and exit the window. In the COMPUTED VAR window, give the variable the name **z.** (The name you give the variable doesn't matter because you will later remove the name from the report.) Adding a character variable is slightly different from adding a numeric variable, which you have done in previous reports. When you add a character variable, you must select the **Character data** check box and, if you want the variable to have a length other than 8, you must type the length in the **Length** field. In this case, use a length of **1.** When your display looks like Display 5.30, select the **Edit Program** push button.

Display 5.30
Adding a
Computed
Character Variable

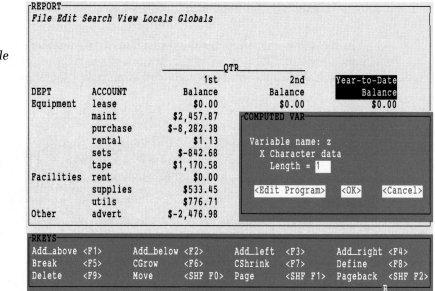

In the COMPUTE window for this new variable, enter the SAS statements in Display 5.31.

Display 5.31
Using Conditional Statements to Define a Computed Variable

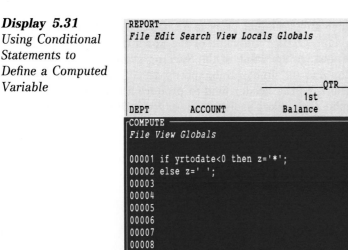

Then exit the COMPUTE and COMPUTED VAR windows. The modified report, which contains an asterisk next to all year-to-date balances that are less than 0, appears in Display 5.32. You may need to issue the RIGHT command (under **View** in the action bar at the top of the REPORT window) to see the new variable.

Display 5.32
Report with New Character Variable Marking Accounts That Are Over Budget

```
┌REPORT─────────────────────────────────────────────────────────────┐
│ File Edit Search View Locals Globals                               │
│                                                                    │
│                                                                    │
│                            ───────QTR───────                       │
│                              1st           2nd      Year-to-Date   │
│ DEPT        ACCOUNT        Balance       Balance        Balance   Z│
│ Equipment   lease           $0.00         $0.00           $0.00    │
│             maint        $2,457.87     $1,324.71       $3,782.58    │
│             purchase    $-8,282.38     $2,230.85      $-6,051.53   *│
│             rental           $1.13       $517.06         $518.19    │
│             sets          $-842.68      $-579.62      $-1,422.30   *│
│             tape         $1,170.58       $573.27       $1,743.85    │
│ Facilities  rent            $0.00         $0.00           $0.00    │
│             supplies      $533.45         $7.52         $540.97    │
│             utils         $776.71        $55.19         $831.90    │
│ Other       advert      $-2,476.98    $-7,325.64      $-9,802.62   *│
├RKEYS───────────────────────────────────────────────────────────────┤
│ Add_above <F1>      Add_below <F2>     Add_left  <F3>   Add_right <F4>│
│ Break     <F5>      CGrow     <F6>     CShrink   <F7>   Define    <F8>│
│○Delete    <F9>      Move      <SHF F0> Page      <SHF F1> Pageback <SHF F2>│
│                                                                  R │
└────────────────────────────────────────────────────────────────────┘
```

Changing the Space between Columns

By default, the REPORT procedure separates columns by the number of spaces specified by the **Spacing** attribute in the ROPTIONS window. In this report, however, you want the asterisk to appear immediately after the number it flags as over budget.

To change the number of spaces that precede an item, select the item and issue the DEFINE command. Change the value of the **Spacing** attribute to suit your needs. In this case, change the value of the **Spacing** attribute in the DEFINITION window for Z to **0**. In addition, specify **$1.** in the **Format** field, and type a blank over the **Z** in the header field to produce a blank column header. When your display looks like Display 5.33, exit the DEFINITION window.

Display 5.33
Closing the Space between Columns

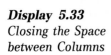

```
┌REPORT────────────────────────────────────────────────────────────┐
│File Edit Search View Locals Globals                               │
│                                                                   │
│       ┌DEFINITION──────────────────────────────────────────────┐  │
│       │                    Definition of Z                     │  │
│ DEPT     A  Usage      Attributes              Options    Color │  │
│ Equipment l  o DISPLAY Format  = $1.        _ NOPRINT    BLUE   │  │
│          m  o ORDER   Spacing = 0          _ NOZERO     RED    │  │
│          p  o GROUP   Width   = 1          _ DESCENDING PINK   │  │
│          r  o ACROSS  Item help =          _ PAGE       GREEN  │  │
│          s  o ANALYSIS Statistic =                      CYAN   │  │
│          t  * COMPUTED                                  YELLOW │  │
│ Facilities r          Type of data      Justification  WHITE  │  │
│          s              o Numeric          o LEFT       ORANGE │  │
│          u              * Character        * RIGHT      BLACK  │  │
│ Other    a                                 o CENTER     MAGENTA│  │
│                                                         GRAY   │  │
│ ┌RKEYS──     Header = [                          ]      BROWN  │  │
│ Add_above <F1                                                  │  │
│ Break    <F5     <Edit Program>   <OK>    <Cancel>             │  │
│ Delete   <F9                                                  │  │
│                                                          R    │  │
└───────────────────────────────────────────────────────────────┘
```

Display 5.34 shows the modified report, with no space between the asterisks and the numbers they flag.

Display 5.34
Report without Space before the Asterisks

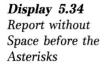

```
┌REPORT────────────────────────────────────────────────────────────┐
│File Edit Search View Locals Globals                               │
│                                                                   │
│                              ──────QTR──────                      │
│                               1st        2nd    Year-to-Date      │
│ DEPT       ACCOUNT          Balance    Balance    Balance         │
│ Equipment  lease             $0.00      $0.00      $0.00          │
│            maint          $2,457.87  $1,324.71  $3,782.58         │
│            purchase      $-8,282.38  $2,230.85 $-6,051.53*        │
│            rental            $1.13    $517.06    $518.19          │
│            sets           $-842.68   $-579.62 $-1,422.30*         │
│            tape          $1,170.58    $573.27  $1,743.85          │
│ Facilities rent              $0.00      $0.00      $0.00          │
│            supplies        $533.45      $7.52    $540.97          │
│            utils           $776.71     $55.19    $831.90          │
│ Other      advert        $-2,476.98 $-7,325.64 $-9,802.62*        │
│                                                                   │
│┌RKEYS─────────────────────────────────────────────────────────── │
│ Add_above <F1>   Add_below <F2>   Add_left  <F3>  Add_right <F4>  │
│ Break     <F5>   CGrow     <F6>   CShrink   <F7>  Define    <F8>  │
│ Delete    <F9>   Move    <SHF F0> Page    <SHF F1> Pageback <SHF F2>
│                                                          R        │
└───────────────────────────────────────────────────────────────┘
```

Changing Column Headers

You are now familiar with a variety of ways to change column headers. For this report you still need to change the following headers:

□ DEPT to Department

□ ACCOUNT to Account

□ QTR to Quarter

The method you choose for changing headers depends on the circumstances. In this case, you can make all the changes simply by typing over the existing headers. When you type over the header for QTR, be sure that one underscore immediately precedes and one underscore immediately follows the word Quarter so that the underscores extend across all columns below QTR.

Issue the REFRESH command to justify all headers. The new report, with revised column headers, appears in Display 5.35.

Display 5.35
Report with
Revised Column
Headers

```
┌REPORT─────────────────────────────────────────────────────────┐
│ File Edit Search View Locals Globals                            │
│                                                                 │
│                                                                 │
│                                   ────────Quarter────────       │
│                                      1st         2nd     Year-to-Date│
│           Department   Account     Balance     Balance     Balance  │
│           Equipment    lease         $0.00       $0.00       $0.00  │
│                        maint      $2,457.87   $1,324.71   $3,782.58 │
│                        purchase  $-8,282.38   $2,230.85  $-6,051.53*│
│                        rental        $1.13     $517.06     $518.19  │
│                        sets       $-842.68    $-579.62  $-1,422.30* │
│                        tape       $1,170.58    $573.27   $1,743.85  │
│           Facilities   rent          $0.00       $0.00       $0.00  │
│                        supplies    $533.45       $7.52     $540.97  │
│                        utils       $776.71      $55.19     $831.90  │
│           Other        advert    $-2,476.98  $-7,325.64  $-9,802.62*│
│                                                                 │
└─────────────────────────────────────────────────────────────────┘
┌RKEYS────────────────────────────────────────────────────────────┐
│ Add_above <F1>     Add_below <F2>    Add_left  <F3>   Add_right <F4>│
│ Break     <F5>     CGrow     <F6>    CShrink   <F7>   Define    <F8>│
│ Delete    <F9>     Move    <SHF F0>  Page    <SHF F1> Pageback <SHF F2>│
│                                                                R   │
└─────────────────────────────────────────────────────────────────┘
```

Adding Break Lines for Each Department

In the previous report you created break lines for each department. You need to do so again for this report. Select **DEPT** as the break variable and issue the BREAK command (under **Edit** in the action bar at the top of the REPORT window). From the LOCATION window select **After detail** to indicate that you want the break lines to appear after the last row for each department, and exit the window. When the BREAK window appears, select

□ **OVERLINE** to separate the break lines from the body of the text with a row of hyphens

□ **PAGE** to create a page break each time the value of DEPT changes

□ **SUMMARIZE** to display summary values for all statistics and computed variables in the report

□ a color of your choice.

   Note: Not all operating systems and devices support all colors, and on some operating systems and devices, one color may map to another color. For example, if the BREAK window displays **BROWN** in yellow characters, selecting **BROWN** results in yellow break lines.

   Currently, color appears only when PROC REPORT displays the report in the REPORT window.

When your display looks like Display 5.36, exit the BREAK window.

Display 5.36
Creating a Break
after Department

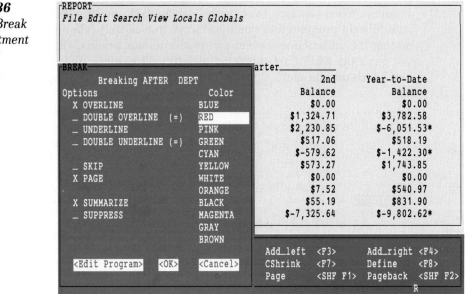

The report now includes a break after DEPT. The first page of the report appears in Display 5.37. You can see the first set of break lines in this display. Use the paging commands described earlier to view the rest of the report.

Display 5.37
Report with a
Break after DEPT

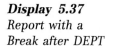

```
┌REPORT──────────────────────────────────────────────────────────┐
│ File Edit Search View Locals Globals                            │
│                                                                 │
│                                                                 │
│                             ───────Quarter───────               │
│                                1st          2nd      Year-to-Date│
│ ▌Department▐ Account         Balance      Balance      Balance   │
│  Equipment   lease            $0.00        $0.00        $0.00    │
│              maint          $2,457.87    $1,324.71    $3,782.58  │
│              purchase      $-8,282.38    $2,230.85   $-6,051.53* │
│              rental           $1.13       $517.06      $518.19   │
│              sets           $-842.68     $-579.62   $-1,422.30*  │
│              tape          $1,170.58      $573.27    $1,743.85   │
│                                                                 │
│  ──────────                ───────────   ───────────  ──────────│
│  Equipment                 $-5,495.48    $4,066.27   $-1,429.21* │
│                                                                 │
│                                                                 │
└─────────────────────────────────────────────────────────────────┘
┌RKEYS────────────────────────────────────────────────────────────┐
│ Add_above <F1>      Add_below <F2>      Add_left  <F3>    Add_right <F4>│
│ Break     <F5>      CGrow     <F6>      CShrink   <F7>    Define    <F8>│
│ Delete    <F9>      Move      <SHF F0>  Page      <SHF F1> Pageback <SHF F2>│
│                                                                  R│
└─────────────────────────────────────────────────────────────────┘
```

When you have viewed the report to your satisfaction, return to the top of the report.

Adding a Customized Break Line

The report in Output 5.2 contains customized break lines at each break. This break line reports the percentage of the year-to-date budget for the entire company allocated to the current department.

Storing Values for Later Use

To calculate the percentage of the budget allocated to each department, you must know the total year-to-date budget for the entire company and the total year-to-date budget for each department. Whenever PROC REPORT creates a break, the summaries of all statistics and computed variables in the report are available in the corresponding report items, whether the break is before or after the detail rows.

For instance, in the current report the values for BUDGET that are to the right of QTR have the following meanings, depending on which row of the report you consider:

□ In a detail row, the value is the year-to-date budget for one account in one department.

□ At a break on DEPT, the value is the year-to-date budget for one department.

□ At a break at the beginning or end of the report, the value is the year-to-date budget for the entire company.

Therefore, to calculate the percentage of the budget allocated to each department, you must create a break at the beginning of the report and store the current value of BUDGET in a new variable, TOTALBUD, whose value does not change as the REPORT procedure builds the report. You then have access to the value for the total budget throughout the report and can use it to calculate the percent of the budget allocated to each department.

So, issue the RBREAK command, which is under **Edit** in the action bar at the top of the REPORT window. Select **Before detail** from the LOCATION window and exit the window. When the BREAK window appears, select the **Edit Program** push button. Enter the DATA step statement that appears in Display 5.38 in the COMPUTE window for this break. Remember that because BUDGET is an analysis variable, you must use a compound name to refer to it.

Display 5.38
Saving a Value Available in a Break for Later Use

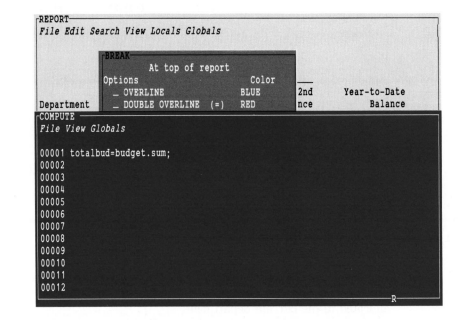

When your display looks like Display 5.38, exit the COMPUTE window and the BREAK window.

Defining and Printing the Customized Break Line

You have already created a break after DEPT. To define the customized break line for that break, you edit the COMPUTE window for that break.

Select **DEPT** and issue the BREAK command. Select **After detail** from the LOCATION window, and select the **Edit Program** push button from the BREAK window.

At the break on department, the year-to-date budget for the current department is in the value of BUDGET. To calculate the percentage of the total budget allocated to the current department, enter the following statements in the COMPUTE window:

```
/* Calculate percentage of total budget allocated */
/* to this department.  Round to the nearest whole */
/* percent.                                         */
pctbud=round((budget.sum/totalbud)*100);
```

You use the LINE statement to print text and variables in a report. The LINE statement provides a subset of the features of the PUT statement, which you use in the DATA step to write lines to the SAS log, the SAS procedure output, or an external file. You must include a format for every variable you display with a LINE statement. For more information on use of the LINE statement, see Chapter 6, "Using the COMPUTE Window," and the documentation of the LINE statement in Chapter 10.

Add the following statements to the COMPUTE window for the break:

```
line ' ';
line pctbud 2.
     '% of the year-to-date budget is allocated to this department.';
```

The first LINE statement writes a blank break line. The second LINE statement writes a break line containing the following elements:

□   the value of PCTBUD using the 2. format

□   the text between quotation marks.

When your display looks like Display 5.39, exit the COMPUTE and BREAK windows.

Display 5.39
Using DATA Step
Statements to
Customize a Break

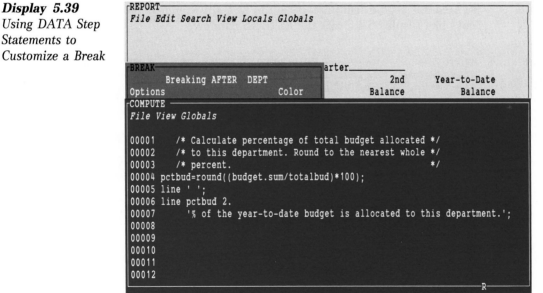

The modified report appears in Display 5.40.

Display 5.40
Report with a
Customized Break

```
┌─REPORT────────────────────────────────────────────────────────────┐
│ File Edit Search View Locals Globals                               │
│                                                                    │
│                                                                    │
│                            ───Quarter───                           │
│                              1st          2nd        Year-to-Date  │
│  Department   Account      Balance      Balance        Balance     │
│  Equipment    lease          $0.00        $0.00          $0.00     │
│               maint      $2,457.87    $1,324.71      $3,782.58     │
│               purchase  $-8,282.38    $2,230.85     $-6,051.53*    │
│               rental         $1.13      $517.06        $518.19     │
│               sets        $-842.68     $-579.62     $-1,422.30*    │
│               tape       $1,170.58      $573.27      $1,743.85     │
│              ──────────   ──────────   ──────────   ──────────────│
│  Equipment               $-5,495.48    $4,066.27     $-1,429.21*   │
│                                                                    │
│  27% of the year-to-date budget is allocated to this department.  │
│                                                                    │
├─RKEYS──────────────────────────────────────────────────────────────┤
│ Add_above <F1>      Add_below <F2>      Add_left  <F3>    Add_right <F4>│
│ Break     <F5>      CGrow     <F6>      CShrink   <F7>    Define    <F8>│
│ Delete    <F9>      Move    <SHF F0>    Page    <SHF F1>  Pageback <SHF F2>│
│                                                                  R │
└────────────────────────────────────────────────────────────────────┘
```

Saving the Report Definition

To save this report definition, issue the RSTORE command. When your display looks like Display 5.41, exit the RSTORE window.

Display 5.41
Saving a Report
Definition

```
┌─REPORT─────────────────────┬─RSTORE─────────────────────────────┐
│ File Edit Search View Locals Globals                            │
│                            │ Description: tutorials: chapter 5(b)│
│                            │ Libname:     sasuser                │
│                            │ Catalog:     tutors                 │
│                       ─────Q│ Report name: chap5b                │
│                          1st│                                    │
│  Department   Account   Balance                    <OK>    <Cancel>│
│  Equipment    lease      $0.00                      │             │
│               maint  $2,457.87                      │             │
│               purchase $-8,282.38     $2,230.85     $-6,051.53*  │
│               rental      $1.13        $517.06        $518.19    │
│               sets     $-842.68       $-579.62     $-1,422.30*   │
│               tape    $1,170.58        $573.27      $1,743.85    │
│              ──────────   ──────────   ──────────   ──────────────│
│  Equipment            $-5,495.48      $4,066.27     $-1,429.21*  │
│                                                                 │
│  27% of the year-to-date budget is allocated to this department.│
│                                                                 │
├─RKEYS────────────────────────────────────────────────────────────┤
│ Add_above <F1>      Add_below <F2>      Add_left  <F3>    Add_right <F4>│
│ Break     <F5>      CGrow     <F6>      CShrink   <F7>    Define    <F8>│
│ Delete    <F9>      Move    <SHF F0>    Page    <SHF F1>  Pageback <SHF F2>│
│                                                                R │
└──────────────────────────────────────────────────────────────────┘
```

Using the REPORT Language to Produce the Report

The following SAS statements produce the first report you developed in this tutorial. For a detailed explanation of these statements, see "Example 4: Using the Same Variable in Multiple Ways" in Chapter 10.

```
proc format;
   value forqtr 1='1st' 2='2nd';
run;

proc report data=report.budget;

   column dept account qtr,budget budget;

   define dept    / group format=$10. width=10 'DEPT';
   define account / group format=$8. width=8 'ACCOUNT';
   define qtr     / across format=forqtr12. width=12 '_QTR_';
   define budget  / sum format=dollar11.2 width=11 'BUDGET';
run;
```

The following statements produce the final report that you produced in this tutorial. For a detailed explanation of these statements, see "Example 5: Creating a Sophisticated Report" in Chapter 10.

```
proc format;
   value forqtr 1='1st' 2='2nd';
run;

options linesize=255 nocenter;

proc report data=report.budget;

   title;

   column dept account qtr,(budget actual qtrbal)
          budget actual yrtodate z;

   define dept    / group format=$10. width=10 'Department';
   define account / group format=$8. width=8 'Account';
   define qtr     / across format=forqtr12. width=12 '_Quarter_';
   define qtrbal  / computed format=dollar11.2 width=11 'Balance';
   define budget  / sum noprint;
   define actual  / sum noprint;
   define yrtodate / computed format=dollar11.2 width=12
                    'Year-to-Date/Balance';
   define z       / computed format=$1. width=1 spacing=0 ' ';

   compute qtrbal;
      _c5_=_c3_-_c4_;
      _c8_=_c6_-_c7_;
   endcomp;
```

```
        compute yrtodate;
           yrtodate=budget.sum-actual.sum;
        endcomp;

        compute z / length=1 char;
           if yrtodate<0 then z='*';
           else z=' ';
        endcomp;

        break after dept / ol page summarize color=red;

        compute after dept;
           pctbud=round((budget.sum/totalbud)*100);
           line ' ';
           line pctbud 2.
                '% of the year-to-date budget is allocated to this department.';
        endcomp;

        compute before;
           totalbud=budget.sum;
        endcomp;
     run;
```

Chapter 6 Using the COMPUTE Window

Introduction

The COMPUTE window enables you to attach DATA step statements and DATA step functions to a break in a report or to a computed variable, and to attach LINE statements to a break in a report. You use these statements to do one of the following:

☐ define a computed variable in the report

☐ calculate values to use in break lines

☐ define break lines that contain text, calculated values, or both.

The REPORT procedure executes these statements as it constructs the report.

You have seen some of the material in this chapter in parts of Chapter 3, "Enhancing Your First Report," Chapter 4, "Grouping and Summarizing Data," and Chapter 5, "Using Some Advanced Features of the REPORT Procedure." However, this chapter consolidates the information on using the COMPUTE window in one place and discusses in detail the principles you need to understand to take advantage of this connection to the DATA step. Specifically, this chapter

□ describes how to attach DATA step statements, DATA step functions, and LINE statements to various parts of a report

□ explains the sequence in which the REPORT procedure executes these statements as it builds the report

□ illustrates different methods of referring to values in a report with DATA step statements

□ lists DATA step statements and functions you can use from the REPORT procedure

□ develops a report that uses multiple COMPUTE windows.

By the time you reach this chapter, you should be comfortable issuing commands, exiting windows, turning DEFER mode on and off, setting characteristics in the DEFINITION window, storing and loading report definitions, and creating breaks on a break variable or at the beginning or end of a report. Earlier chapters describe how to do all these things. Detailed directions are no longer provided.

Note: The syntactical equivalent of the COMPUTE window in the nonwindowing environment is the COMPUTE statement. For details on the use of the COMPUTE statement, refer to "COMPUTE Statement" in Chapter 10, "The REPORT Language."

Understanding the Details of Operation

In order to use the COMPUTE window successfully, you must understand when PROC REPORT executes the statements you place in the COMPUTE window.

Building a Report

The first thing PROC REPORT does when it builds a report is to consolidate the data by group or order variables. It then calculates all statistics you use in the report, those for detail rows as well as those for summary lines in breaks. Once the REPORT procedure has calculated statistics, it begins constructing the rows of the report. At the beginning of each row, it initializes all columns to missing. Whenever it comes to a column that contains values for a computed variable, PROC REPORT executes the statements in the COMPUTE window attached to that variable. Whenever it comes to a break, it first constructs the break lines created in the BREAK window (for example, overlining, underlining, and a summary line). It then executes DATA step statements attached to the break. As you will see shortly, these statements can create customized break lines or alter values in the summary line.

The statements in a COMPUTE window are a *code segment.*

Consider, for example, the report in Output 6.1. In this report DEPT is a group variable; BUDGET and ACTUAL are analysis variables used to calculate the SUM statistic; BALANCE is a computed variable; and 'Number of Observations' is the N statistic, which indicates how many observations each detail row represents. At the end of the report is a break that includes summaries of the statistics and computed variables in the report as well as the value of **TOTAL** for DEPT.

Output 6.1
Report Used to Illustrate the Process of Building a Report

```
                                                                        1


                     Amount        Amount              Number of
        Department   Budgeted      Spent      Balance  Observations
        ------------------------------------------------------------

        Equipment    $207,000.00  $208,429.21  $-1,429.21          12
        Facilities    $62,000.00   $60,627.13   $1,372.87           6
        Other        $101,000.00  $108,640.44  $-7,640.44           6
        Staff        $395,000.00  $393,857.51   $1,142.49           4
        Travel        $10,000.00    $8,456.99   $1,543.01           4
        ===========  ===========  ===========  ===========  ============
        TOTAL        $775,000.00  $780,011.28  $-5,011.28          32
        ===========  ===========  ===========  ===========  ============
```

PROC REPORT starts building the report by consolidating the data (DEPT, in the column labeled 'Department', is a group variable) and calculating the statistics (BUDGET.SUM and ACTUAL.SUM in the columns labeled 'Amount Budgeted' and 'Amount Spent') for each detail row and for the break. Now, the REPORT procedure is ready to start building the first row of the report. A description of the process follows:

1. PROC REPORT initializes all report variables to missing. Figure 6.1 illustrates the first detail row at this point. Missing values for a character variable are represented by a blank, and missing values for a numeric variable are represented by a period.

Figure 6.1
Initializing Values for the First Detail Row

Department	Amount Budgeted	Amount Spent	Balance	Number of Observations
	.	.	.	.

2. Figure 6.2 illustrates the construction of the first three columns of the row. PROC REPORT fills in values for the row from left to right.

Figure 6.2
Filling in Values from Left to Right

Department	Amount Budgeted	Amount Spent	Balance	Number of Observations
Equipment	.	.	.	.

Figure 6.2
(continued)

Department	Amount Budgeted	Amount Spent	Balance	Number of Observations
Equipment	$207,000.00	.	.	.

Department	Amount Budgeted	Amount Spent	Balance	Number of Observations
Equipment	$207,000.00	$208,429.21	.	.

3. The next column in the report contains the computed variable BALANCE. When it gets to this column, PROC REPORT executes the statement in the COMPUTE window attached to the variable. This statement simply calculates the difference between BUDGET and ACTUAL:

```
balance=budget.sum-actual.sum;
```

The row now looks like Figure 6.3.

Figure 6.3
Adding a
Computed Value
to the Row

Department	Amount Budgeted	Amount Spent	Balance	Number of Observations
Equipment	$207,000.00	$208,429.21	$-1,429.21	.

4. Next, PROC REPORT fills in the value for the N statistic and writes the row to the report. Figure 6.4 illustrates the completed row.

Figure 6.4
First Complete
Detail Row

Department	Amount Budgeted	Amount Spent	Balance	Number of Observations
Equipment	$207,000.00	$208,429.21	$-1,429.21	12

5. PROC REPORT repeats steps 1, 2, 3, and 4 for each detail row in the report.

6. When PROC REPORT gets to the break at the end of the report, it constructs a preliminary version of the summary line in the same way it constructed each detail line. The REPORT procedure creates this preliminary version of the summary line before it executes any statements in the COMPUTE window attached to the break. Figure 6.5 illustrates this version of the summary line.

Figure 6.5
Preliminary
Summary Line

Department	Amount Budgeted	Amount Spent	Balance	Number of Observations
	$775,000.00	$780,011.28	$-5,011.28	32

7. If no COMPUTE window is attached to the break, the preliminary version of the summary line is the same as the final version. However, in the example being described, a COMPUTE window is attached to the break. Therefore, the next thing that PROC REPORT does is execute the statements in that COMPUTE window. In this case, the COMPUTE window contains the following statement:

```
dept='TOTAL';
```

This statement replaces the value of DEPT, which in the summary line is missing by default, with the word **TOTAL**. After PROC REPORT executes the statement, it modifies the summary line to reflect this change to the value of DEPT. The final version of the summary line appears in Figure 6.6.

Figure 6.6
Final Summary
Line

Department	Amount Budgeted	Amount Spent	Balance	Number of Observations
TOTAL	$775,000.00	$780,011.28	$-5,011.28	32

Finally, PROC REPORT writes the last summary line and the two break lines that visually separate it from the rest of the report.

Initializing Variables

Variables that appear only in one or more COMPUTE windows are *DATA step variables*. Variables that may appear in one or more COMPUTE windows but also appear in one or more columns of the report are *report variables*. Whereas PROC REPORT initializes report variables at the beginning of each row of the report, the value for a DATA step variable is initialized to 0 (for a numeric variable) or blank (for a character variable) and remains 0 or blank until you specifically assign a value. PROC REPORT retains the value of a DATA step variable from the execution of one code segment to another.

Because all code segments share the current values of all variables, you can initialize DATA step variables at a break at the beginning of the report or at a break before a break variable. Several examples of initializing appear later in this chapter.

Referring to Values in a Report

When you use DATA step statements to reference the current value of an item in a report, the form of the name you use depends on the way you are using the item in the report. The following rules apply when the value you are referencing is not in a column below an across variable:

□ If the value is the value of a group variable, an order variable, a display variable that has no statistic above or below it, or a computed variable, refer to the value by the name of the variable. For example, the following statement creates the computed variable BALANCE and assigns it the difference between BUDGET and ACTUAL. In this case BUDGET and ACTUAL are display variables without any statistics above or below them.

```
balance=budget-actual;
```

You used this form of accessing values in "Adding a Computed Variable" in Chapter 3.

□ If the value you are referencing is the value of an analysis variable or a display variable with a statistic above or below it, refer to the value by a compound name that includes the name of the variable and the name of the statistic, separated by a period (.). For example, the following statement creates the computed variable BALANCE and assigns it the difference between BUDGET and ACTUAL when these variables are either display variables with the SUM statistic above or below them or analysis variables used to calculate the SUM statistic:

```
balance=budget.sum-actual.sum;
```

You used this form of accessing values in "Computing a Variable Based on Statistics" in Chapter 4.

If the value you want to reference is in a column below an across variable, you must use the following form to refer to the value:

_Cn_

where *n* is the number of the column containing the value. For example, the following statement defines the value in a new column of the report (in the fifth column) as the difference between the values in the third and fourth columns:

```
_c5_=_c3_-_c4_;
```

You used this form of accessing values in "Referring to Report Items by Column Number" in Chapter 5.

Using DATA Step Statements and Functions

You can use the following DATA step statements in code segments:

CALL	%INCLUDE
DM	LENGTH
DO (all forms)	LINK
END	RETURN
GO TO	SELECT
IF-THEN/ELSE	

You can also use assignment, comment, null, and sum statements. You can use all DATA step functions in code segments.

Developing a Report Using Multiple COMPUTE Windows

Now that you have an understanding of the fundamentals of using the COMPUTE window, you are ready to begin creating the report in Output 6.2.

Output 6.2
Using Multiple
COMPUTE
Windows to Create
a Report

```
                                                                        1

     Quarter  Department      Balance       Budget       Actual

        1  Equipment      $-5,495.48  $109,500.00  $114,995.48
           Facilities      $1,310.16   $31,750.00   $30,439.84
           Other          $-1,514.21   $46,500.00   $48,014.21
           Staff          $-1,492.80  $170,000.00  $171,492.80
           Travel            $717.59    $4,300.00    $3,582.41

The largest overdraw for this quarter was in the Equipment department.
It was overdrawn by $5,495.48.

        2  Equipment       $4,066.27   $97,500.00   $93,433.73
           Facilities         $62.71   $30,250.00   $30,187.29
           Other          $-6,126.23   $54,500.00   $60,626.23
           Staff           $2,635.29  $225,000.00  $222,364.71
           Travel            $825.42    $5,700.00    $4,874.58

The largest overdraw for this quarter was in the Other department.
It was overdrawn by $6,126.23.

           ----------    -----------  -----------  -----------
           Total:        $-5,011.28  $775,000.00  $780,011.28
           ----------    -----------  -----------  -----------

========================================================================
The largest overdraw was in the Other department during the 2nd quarter.
It was overdrawn by $6,126.23.
========================================================================
```

This report uses COMPUTE windows attached to each of the following:

□ a break at the beginning of the report to initialize DATA step variables

□ a break before QTR to store the value of QTR so you can use it in customized break lines and to initialize DATA step variables that you need to initialize each time the value of QTR changes

□ the variable BALANCE to compute the value of BALANCE and to perform calculations necessary for the customized break lines

□ a break after QTR to print customized break lines for each quarter

□ a break at the end of the report to print customized break lines for the entire report.

The remainder of this chapter is devoted to the step-by-step development of this report. The text assumes that you are familiar with simple SAS text editor commands, such as the commands that insert and delete lines of text. If you need information on these commands, consult Chapter 19, "SAS Text Editor Commands," in *SAS Language: Reference, Version 6, First Edition.*

The first thing you need to do is to invoke the REPORT procedure with the following statements. The FORQTR. format is used in some of the COMPUTE windows to show the value of QTR as 1st or 2nd rather than as 1 or 2.

```
proc format;
   value forqtr 1='1st' 2='2nd';
run;


title;


proc report data=report.budget windows ps=60;
   column qtr dept budget actual;
run;
```

Notice the PS= option, which sets the page size of the report to 60 lines per page. This enables you to view the entire report as you develop it using only scrolling commands. Paging will not be necessary.

Your display now looks like Display 6.1.

Display 6.1
Initial Report

```
┌REPORT─────────────────────────────────────────────────────────────┐
│ File Edit Search View Locals Globals                               │
│                                                                  1 │
│                                                                    │
│                  QTR  DEPT            BUDGET        ACTUAL          │
│                    1  Staff      $130,000.00   $127,642.68         │
│                    2  Staff      $165,000.00   $166,345.75         │
│                    1  Staff       $40,000.00    $43,850.12         │
│                    2  Staff       $60,000.00    $56,018.96         │
│                    1  Equipment   $40,000.00    $40,000.00         │
│                    2  Equipment   $40,000.00    $40,000.00         │
│                    1  Equipment   $40,000.00    $48,282.38         │
│                    2  Equipment   $20,000.00    $17,769.15         │
│                    1  Equipment    $8,000.00     $6,829.42         │
│                    2  Equipment   $12,000.00    $11,426.73         │
│                    1  Equipment    $7,500.00     $8,342.68         │
│                    2  Equipment    $7,500.00     $8,079.62         │
├RKEYS──────────────────────────────────────────────────────────────┤
│ Add_above <F1>     Add_below <F2>     Add_left  <F3>     Add_right <F4> │
│ Break     <F5>     CGrow     <F6>     CShrink   <F7>     Define    <F8> │
│ Delete    <F9>     Move    <SHF F0>   Page    <SHF F1>   Pageback <SHF F2> │
│                                                            R        │
└────────────────────────────────────────────────────────────────────┘
```

Placing a Computed Variable in the Report

Because PROC REPORT constructs each row of the report from left to right, a computed variable can rely only on values that appear to its left in the report. For example, select **Balance** and use the MOVE command (under **Edit** in the action bar at the top of the REPORT window) to move it from its current position to the left of BUDGET. When you do so, the REPORT procedure issues the following message in the MESSAGE window:

```
NOTE: Missing values were generated as a result of performing an
      operation on missing values.
```

When you exit the MESSAGE window, your display looks like Display 6.5. The column for BALANCE contains only periods because the values for BUDGET and ACTUAL are missing when PROC REPORT executes the statements in the COMPUTE window. (Recall that PROC REPORT initializes report variables to missing at the beginning of each row, so the values of these variables are missing for any computations that occur to their left.)

Display 6.5
Missing Values for
a Computed
Variable

```
┌REPORT─────────────────────────────────────────────────────────────┐
│File Edit Search View Locals Globals                                │
│                                                                   1│
│                                                                    │
│         Quarter  Department    Balance   ▐ BUDGET    ACTUAL         │
│               1  Equipment         .   $109,500.00 $114,995.48      │
│                  Facilities        .    $31,750.00  $30,439.84      │
│                  Other             .    $46,500.00  $48,014.21      │
│                  Staff             .   $170,000.00 $171,492.80      │
│                  Travel            .     $4,300.00   $3,582.41      │
│               2  Equipment         .    $97,500.00  $93,433.73      │
│                  Facilities        .    $30,250.00  $30,187.29      │
│                  Other             .    $54,500.00  $60,626.23      │
│                  Staff             .   $225,000.00 $222,364.71      │
│                  Travel            .     $5,700.00   $4,874.58      │
│                                                                    │
└────────────────────────────────────────────────────────────────────┘
┌RKEYS───────────────────────────────────────────────────────────────┐
│ Add_above <F1>      Add_below <F2>      Add_left  <F3>    Add_right <F4> │
│ Break     <F5>      CGrow     <F6>      CShrink   <F7>    Define    <F8> │
│ Delete    <F9>      Move      <SHF F0>  Page      <SHF F1> Pageback <SHF F2> │
│                                                                  R │
└────────────────────────────────────────────────────────────────────┘
```

If you want a computed variable to appear to the left of the variables you use to compute it, you must include those variables to the left and right of the computed variable and suppress the left-hand display with the NOPRINT option in the DEFINITION window. To do so, first simultaneously add BUDGET and ACTUAL to the left of BALANCE by selecting **Balance**, issuing the ADD_LEFT command, selecting **Dataset variable** from the ADDING window, and selecting both **BUDGET** and **ACTUAL** from the DATASET VAR window. In the DEFINITION windows for BUDGET and ACTUAL, which PROC REPORT automatically displays one after the other, accept the default usage. The first page of the report appears in Display 6.6. BUDGET and ACTUAL now appear both to the left and right of BALANCE.

Display 6.6
Adding BUDGET
and ACTUAL to
the Report a
Second Time

```
┌REPORT─────────────────────────────────────────────────────────────┐
│File Edit Search View Locals Globals                                │
│NOTE: More columns appear on the next page.                         │
│                                                                 1  │
│                                                                    │
│    Quarter  Department       BUDGET        ACTUAL     ┌Balance┐      BUDGET│
│          1  Equipment   $109,500.00   $114,995.48   $-5,495.48  $109,500.00│
│             Facilities   $31,750.00    $30,439.84    $1,310.16   $31,750.00│
│             Other        $46,500.00    $48,014.21   $-1,514.21   $46,500.00│
│             Staff       $170,000.00   $171,492.80   $-1,492.80  $170,000.00│
│             Travel        $4,300.00     $3,582.41      $717.59    $4,300.00│
│          2  Equipment    $97,500.00    $93,433.73    $4,066.27   $97,500.00│
│             Facilities   $30,250.00    $30,187.29       $62.71   $30,250.00│
│             Other        $54,500.00    $60,626.23   $-6,126.23   $54,500.00│
│             Staff       $225,000.00   $222,364.71    $2,635.29  $225,000.00│
│             Travel        $5,700.00     $4,874.58      $825.42    $5,700.00│
│                                                                    │
│                                                                    │
└────────────────────────────────────────────────────────────────────┘
┌RKEYS───────────────────────────────────────────────────────────────┐
│Add_above <F1>      Add_below <F2>      Add_left  <F3>      Add_right <F4>│
│Break     <F5>      CGrow     <F6>      CShrink   <F7>      Define    <F8>│
│Delete    <F9>      Move    <SHF F0>    Page    <SHF F1>    Pageback <SHF F2>│
│                                                                  R │
└────────────────────────────────────────────────────────────────────┘
```

Now suppress the display of the first appearances of BUDGET and ACTUAL by selecting **NOPRINT** in their DEFINITION windows. In the modified report, which appears in Display 6.7, BALANCE appears to be to the left of BUDGET and ACTUAL.

Display 6.7
Report with
Computed Variable
Appearing to the
Left of the
Variables Used to
Compute It

```
┌REPORT─────────────────────────────────────────────────────────────┐
│File Edit Search View Locals Globals                                │
│                                                                    │
│                                                                 1  │
│                                                                    │
│        ┌Quarter┐ Department       Balance        BUDGET        ACTUAL│
│              1  Equipment    $-5,495.48   $109,500.00   $114,995.48│
│                 Facilities    $1,310.16    $31,750.00    $30,439.84│
│                 Other        $-1,514.21    $46,500.00    $48,014.21│
│                 Staff        $-1,492.80   $170 ,000.00  $171,492.80│
│                 Travel          $717.59     $4,300.00     $3,582.41│
│              2  Equipment     $4,066.27    $97,500.00    $93,433.73│
│                 Facilities       $62.71    $30,250.00    $30,187.29│
│                 Other        $-6,126.23    $54,500.00    $60,626.23│
│                 Staff         $2,635.29   $225,000.00   $222,364.71│
│                 Travel          $825.42     $5,700.00     $4,874.58│
│                                                                    │
│                                                                    │
└────────────────────────────────────────────────────────────────────┘
┌RKEYS───────────────────────────────────────────────────────────────┐
│Add_above <F1>      Add_below <F2>      Add_left  <F3>      Add_right <F4>│
│Break     <F5>      CGrow     <F6>      CShrink   <F7>      Define    <F8>│
│Delete    <F9>      Move    <SHF F0>    Page    <SHF F1>    Pageback <SHF F2>│
│                                                                  R │
└────────────────────────────────────────────────────────────────────┘
```

Customizing a Summary Line

By default the summary line at the end of a report contains values only for the statistics and computed variables in the report. However, the summary line at the end of the report you are working on is customized to contain the value `Total:` in the column labeled 'Department'. You customize the summary line with DATA step statements attached to the same break that creates the summary line.

Now, use the RBREAK command to create a break at the end of the report. From the BREAK window select **OVERLINE, UNDERLINE, SUMMARIZE,** and a color of your choice to create the default summary line.

Note: Not all operating systems and devices support all colors, and on some operating systems and devices, one color may map to another color. For example, if the BREAK window displays **BROWN** in yellow characters, selecting **BROWN** results in yellow break lines.

Next, select the **Edit Program** push button to open the COMPUTE window attached to this break.

In the COMPUTE window enter the following assignment statement:

```
dept='Total:';
```

Then exit the COMPUTE and BREAK windows. You can see the customized break line in Display 6.8.

Display 6.8
Customizing a
Summary Line

```
┌REPORT─────────────────────────────────────────────────────────────────┐
│ File Edit Search View Locals Globals                                   │
│                                                                      1 │
│                                                                        │
│         ▐Quarter▌ Department        Balance       BUDGET       ACTUAL  │
│               1  Equipment      $-5,495.48  $109,500.00  $114,995.48   │
│                  Facilities      $1,310.16   $31,750.00   $30,439.84   │
│                  Other          $-1,514.21   $46,500.00   $48,014.21   │
│                  Staff          $-1,492.80  $170,000.00  $171,492.80   │
│                  Travel            $717.59    $4,300.00    $3,582.41   │
│               2  Equipment       $4,066.27   $97,500.00   $93,433.73   │
│                  Facilities         $62.71   $30,250.00   $30,187.29   │
│                  Other          $-6,126.23   $54,500.00   $60,626.23   │
│                  Staff           $2,635.29  $225,000.00  $222,364.71   │
│                  Travel            $825.42    $5,700.00    $4,874.58   │
│                                 -----------  -----------  -----------  │
│                  Total:         $-5,011.28  $775,000.00  $780,011.28   │
│                                                                        │
└────────────────────────────────────────────────────────────────────────┘
┌RKEYS───────────────────────────────────────────────────────────────────┐
│ Add_above <F1>      Add_below <F2>      Add_left  <F3>   Add_right <F4>  │
│ Break     <F5>      CGrow     <F6>      CShrink   <F7>   Define    <F8>  │
│ Delete    <F9>      Move      <SHF F0>  Page      <SHF F1>  Pageback <SHF F2> │
│                                                                      R │
└────────────────────────────────────────────────────────────────────────┘
```

Calculating Values for a Set of Rows

The report you are building (which appears in Output 6.1) shows the figures for BUDGET, ACTUAL, and BALANCE for each department for each quarter. The customized break lines at the breaks after QTR report which department overspent its budget the most during each quarter. The customized break lines at the end of the report state which department has overspent its budget the most for the year to date. You will first modify the report to produce the customized break lines at the end of the report.

In order to produce these break lines, you must determine in which detail row of the report the value of BALANCE is at a minimum. To do so, you use two DATA step variables. The first one, MINBAL, stores the minimum value of BALANCE encountered up to the current row of the report. The other, MINDEPT, stores the name of the corresponding department. Before creating the first detail row of the report, you need to initialize MINBAL and MINDEPT. Because you need to initialize these variables only once, create a break at the beginning of the report and initialize the variables in the COMPUTE window attached to that break. To create this break, issue the RBREAK command and select **Before detail** from the LOCATION window. From the BREAK window, select the **Edit Program** push button. The COMPUTE window now appears on your display. Enter the assignment statements that initialize MINBAL and MINDEPT. When your display looks like Display 6.9, exit the COMPUTE and BREAK windows.

Display 6.9
Initializing
Variables at a
Break at the
Beginning of the
Report

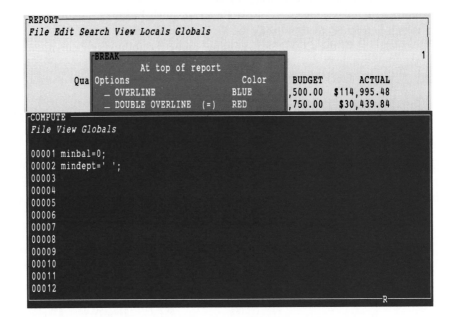

PROC REPORT executes these statements only once, just before it starts constructing the first detail row of the report.

Note: By default, PROC REPORT initializes numeric DATA step variables to 0 and character DATA step variables to a blank. These values remain in effect until you change them. In this case, the values you use to initialize MINBAL and MINDEPT are the defaults. Therefore, it was not really necessary to explicitly initialize the variables. The process is shown here so that you'll understand how to initialize variables when you want to use values other than the defaults.

Now that you have initialized MINBAL and MINDEPT, you are ready to determine which detail row contains the minimum balance. After calculating BALANCE for the current row, you must compare the value of BALANCE to MINBAL. If BALANCE is greater, do nothing. If BALANCE is less than MINBAL, the current department has overspent its budget by more than any other department encountered so far. In this case, store the value of BALANCE in MINBAL and the value of DEPT in MINDEPT. After you have processed the last detail row in the report, the value in MINDEPT is the name of the department that overspent its budget by the most, and the value in MINBAL is the amount by which that department overspent its budget. Because you must perform the comparison of BALANCE to MINBAL on every detail row after PROC REPORT has filled in the value for BALANCE, place the statements that do the comparison in a COMPUTE window associated with the computed variable BALANCE.

Select **Balance**, issue the DEFINE command, and select the **Edit Program** push button from the DEFINITION window. In the COMPUTE window that appears, enter below the assignment statement that defines BALANCE the statements that appear in Display 6.10.

Display 6.10
Calculating
MINBAL and
MINDEPT

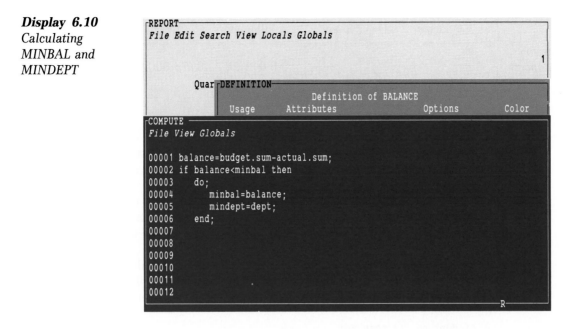

```
┌REPORT─────────────────────────────────────────────────────────┐
│File Edit Search View Locals Globals                            │
│                                                                │
│                                                             1  │
│         Quar┌DEFINITION─────────────────────────────────────┐  │
│             │                   Definition of BALANCE        │
│             │     Usage     Attributes         Options    Color│
┌COMPUTE ─────┴──────────────────────────────────────────────────
│File View Globals
│
│00001 balance=budget.sum-actual.sum;
│00002 if balance<minbal then
│00003    do;
│00004       minbal=balance;
│00005       mindept=dept;
│00006    end;
│00007
│00008
│00009
│00010
│00011       .
│00012
│                                                          ─R─
└──────────────────────────────────────────────────────────────
```

The IF-THEN statement determines which row of the report contains the minimum value of BALANCE. The assignment statements in the DO group assign the value of the minimum balance to MINBAL and the name of the corresponding department to MINDEPT. PROC REPORT executes this code segment each time it reaches the column for the variable BALANCE, that is, once for each detail row and once for each summary line of the report. However, in this case, you do not want the IF-THEN statement (and the DO group associated with it) to execute on the summary lines of the report. If the negative balances were distributed more evenly over the company, it would be possible for the value of BALANCE that appears in the summary line to be less than the BALANCE for any individual department. In such a case, the current report would consider the balance in the summary line to be the minimum. Since you want to determine the individual department with the lowest value of BALANCE, you must prevent PROC REPORT from executing these statements on the summary line.

Preventing Execution of Statements on a Summary Line

In order to prevent PROC REPORT from executing statements on the summary line of the report, you must be able to distinguish the summary line from the detail rows. The value of DEPT in the summary line in this report is `Total:`. Because this value appears in no other row of the report, you might expect that you could distinguish the summary line from all other rows of the report by testing the value of DEPT in each row and preventing execution of the appropriate statements when the value of DEPT is `Total:`.

However, you must remember the order in which the REPORT procedure constructs a report. The statements whose execution you want to prevent are in the COMPUTE window attached to the computed variable BALANCE. Recall that when PROC REPORT gets to a break at the end of a report, it constructs a preliminary version of the summary line before it executes any code in the COMPUTE window attached to the break. In this case, the preliminary summary line looks like Figure 6.7.

Figure 6.7
Preliminary
Summary Line for
the Current Report

Quarter	Department	Balance	BUDGET	ACTUAL
		$-5,011.28	$775.000.00	$780,011.29

It is during the construction of this preliminary summary line, when it gets to the column for BALANCE, that PROC REPORT executes the statements in the COMPUTE window attached to BALANCE. At this point, the value of DEPT is a blank. It does not change to `Total:` until PROC REPORT executes the statements in the COMPUTE window attached to the break at the end of the report. No other row of the report has a blank for the value of DEPT. Therefore, you can distinguish detail rows from the summary line by the fact that on detail rows the value of DEPT is not blank.

To execute the statements that determine the minimum value of BALANCE only if the current value of DEPT is not missing, insert the following statement between the definition of BALANCE and the IF-THEN statement:

```
if dept ne ' ' then do;
```

Enter an END statement at the end of the code segment:

```
end;
```

When your display looks like Display 6.11, exit the COMPUTE and DEFINITION windows.

Display 6.11
Preventing
Execution on a
Summary Line

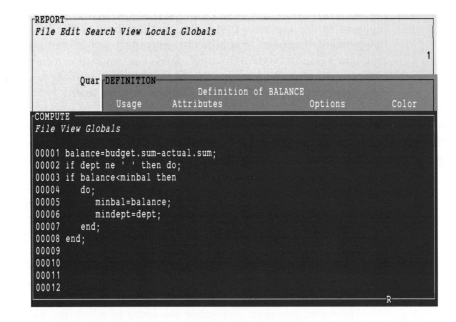

Note: The appearance of your report doesn't change because although you have now calculated values for MINBAL and MINDEPT, you have not yet written these values anywhere in the report.

Writing Customized Break Lines

After PROC REPORT processes the last detail row of the report, it has the values you want to display in the customized break line. To display these values in a break line, you must use the LINE statement in a COMPUTE window attached to a break. In this case, the break is a break at the end of the report.

Understanding the LINE Statement

The LINE statement writes break lines containing text, values calculated for a set of rows of the report, or both. You can use it in the COMPUTE window associated with a break on a variable or a break at the beginning or end of a report. Each LINE statement writes one customized break line. Characters in quotation marks in the LINE statement appear in the break line. Other characters generally specify variables and their formats. For a complete discussion of the LINE statement, see Chapter 10.

It is important to realize that the LINE statement does not participate in the flow of execution through a code segment. In other words, LINE statements take effect only after PROC REPORT has executed all other executable statements in the code segment. This fact has several repercussions:

□ If you alter the value of a variable in the same code segment in which you use it in a LINE statement, the LINE statement uses the last value you assign to the variable in the code segment, regardless of the location of the assignment within the code segment. For example, consider the following code segment:

```
x=1;
line 'The value of X is ' x 1.;
x=0;
```

These statements always print the text "The value of X is 0" because the LINE statement uses the values that exist after PROC REPORT has executed all other executable statements (in this case both assignment statements) in the code segment.

□ You cannot use the LINE statement in conditional statements (IF-THEN and IF-THEN/ELSE statements).

□ You cannot use the LINE statement in iterative loops (created with the DO statement).

Using the LINE Statement to Create Customized Break Lines

Now you must modify the break at the end of the report to print the customized break lines you want. Issue the RBREAK command, select **After detail** from the LOCATION window, and, in the BREAK window, select the **Edit Program** push button. In the COMPUTE window that appears, type the code segment for the customized break lines. This code segment consists of seven statements. The first two statements and the last statement simply improve the appearance of the report. The first statement, which follows, writes a blank line that separates the customized break lines from the noncustomized break lines:

```
line ' ';
```

The second and the last statements are identical:

```
line 75*'=';
```

This statement uses a feature of the LINE statement that enables you to write repetitions of one or more characters. You specify the number of repetitions, followed by an asterisk and the string (in quotation marks) that you want repeated. For example, each occurrence of this statement writes a line of 75 equals signs (=) to the report.

The third statement creates a new variable, LEN, which is the length of the current value of MINDEPT:

```
len=length(mindept);
```

You calculate this variable in order to use the $VARYING. format, which enables you to print the value of MINDEPT in the appropriate number of horizontal positions (for example, 9 for `Equipment` but only 5 for `Other`).

The fourth statement creates a variable whose value is the absolute value of MINBAL so that you can print a positive number in the report:

```
absmin=abs(minbal);
```

The fifth and sixth statements print the customized break lines containing the information about the department with the lowest balance:

```
line 'The largest overdraw was in the ' mindept $varying. len
    ' department.';
line 'It was overdrawn by ' absmin dollar9.2 '.';
```

In addition to the text in quotation marks, these statements print the value of MINDEPT with the $VARYING. format and the value of ABSMIN with the DOLLAR9.2 format. For information on the $VARYING. format, see Chapter 14, "SAS Formats," in *SAS Language: Reference*.

Now enter all seven statements in the COMPUTE window attached to the break at the end of your report. When your display looks like Display 6.12, exit the COMPUTE and BREAK windows.

Display 6.12
Using the LINE
Statement to
Create Customized
Break Lines

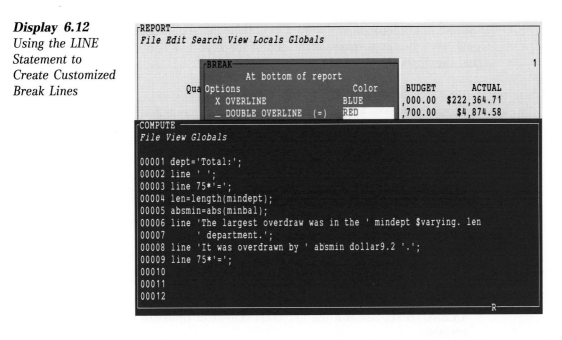

Your current report now contains a break at the end of the report that includes customized break lines. However, you may need to scroll through your report to see the break lines, which appear in Display 6.13.

Display 6.13
Customized Break
Lines at the End of
the Report

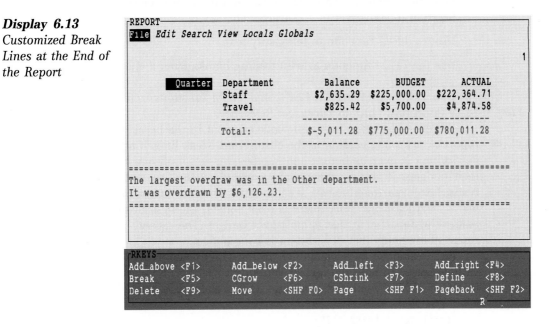

Conditionally Constructing Break Lines

As you learned earlier, because the LINE statement doesn't participate in the flow of execution through a code segment, you cannot use the LINE statement in an IF-THEN/ELSE construction. However, you may want to vary the content of your break lines depending on the values in your report. For example, the current report definition works fine as long as some department is over its budget.

However, if all departments were within their budgets, MINBAL and MINDEPT would have the same values at the end of the report as they do at the beginning. Thus, the break lines would read as follows:

```
The largest overdraw was in the   department.
It was overdrawn by        $0.00.
```

To avoid this situation, you may want to construct an alternate break line that says "No departments were over budget." to print in the happy event that all departments are within their budgets.

When you need to construct alternate text for your customized break lines, you do not print the text directly. Rather, you use conditional SAS statements to store different versions of the text in a character variable. Then you use the LINE statement to print that character variable.

When you construct character variables this way, you must be aware of several things:

□ A character variable takes its length from the first value assigned to it or from a LENGTH statement if the LENGTH statement precedes the first use of the variable.

□ The concatenation operator (||) joins character values.

□ You can concatenate only character values. Therefore, to get the value of a numeric variable, such as MINBAL, into the character string, use the PUT function to convert it to a character string.

□ Use the LEFT function to left-justify text.

□ Use the TRIM function to trim trailing blanks from a character string.

For information on the PUT, LEFT, and TRIM functions, refer to Chapter 11, "SAS Functions," in *SAS Language: Reference.*

For example, let's change the code segment in the break at the end of this report to print one version of the customized break lines if at least one department is over its budget and another version if all departments are within their budgets. If no departments are overbudget, the first customized break line reads, "No departments were over budget." The second customized break line is blank. If at least one department is over budget, the first customized break line names the department that is most over budget. The second customized break line reports how much that department is over budget.

First, establish an adequate length for TEXT1 and TEXT2, the two character variables used to hold the text. If you didn't use the LENGTH statement, the length of TEXT1 would be 1 because its first value is a single blank.

```
length text1 text2 $ 75;
```

Next, use conditional logic to assign the appropriate values to TEXT1 and
TEXT2:

```
if minbal=0 then
   do;
      text1='No departments were over budget.';
      text2=' ';
   end;
else
   do;
      text1='The largest overdraw was in the ' || trim(mindept)
            || ' department.';
      text2='It was overdrawn by '
            || trim(left(put(abs(minbal),dollar11.2))) || '.';
   end;
```

The last assignment statement (for TEXT2) contains four nested SAS functions:
TRIM, LEFT, PUT, and ABS. The ABS function takes the absolute value of
MINBAL. The PUT function converts the absolute value of MINBAL with a format
of DOLLAR11.2 to a character string. The LEFT function left-aligns the resulting
character string, and the TRIM function removes the trailing blanks.

Finally, write the character variables in the customized break lines using the
$75. format. The actual content of TEXT1 and TEXT2 depends on whether or not
at least one department is over its budget.

```
line ' ';
line 75*'=';
line text1 $75.;
line text2 $75.;
line 75*'=';
```

Put the entire code segment in the COMPUTE window attached to the break
at the end of the report by issuing the RBREAK command, selecting **After detail**
from the LOCATION window, selecting the **Edit Program** push button from the
BREAK window, and replacing all but the first statement with the new set of
statements. When your display looks like Display 6.14, exit the COMPUTE and
BREAK windows.

Display 6.14
*Using Conditional
Logic to Customize
Break Lines*

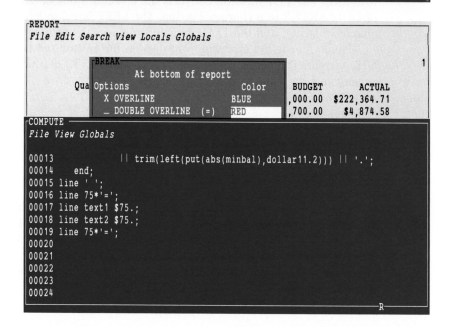

```
┌REPORT──────────────────────────────────────────────────────────┐
│File Edit Search View Locals Globals                             │
│                                                              1  │
│         ┌BREAK─────────────────────────────┐                    │
│         │        At bottom of report       │                    │
│   Qua│ Options                     Color   │  BUDGET      ACTUAL │
│         │   X OVERLINE                BLUE   │  ,000.00  $222,364.71│
│         │   _ DOUBLE OVERLINE    (=) │RED    │  ,700.00   $4,874.58│
┌COMPUTE ─────────────────────────────────────────────────────────┐
│File View Globals                                                │
│                                                                 │
│00001 dept='Total:';                                             │
│00002 length text1 text2 $ 75;                                   │
│00003 if minbal=0 then                                           │
│00004    do;                                                     │
│00005        text1='No departments were over budget.';           │
│00006        text2=' ';                                          │
│00007    end;                                                    │
│00008 else                                                       │
│00009    do;                                                     │
│00010        text1='The largest overdraw was in the ' || trim(mindept)│
│00011            || ' department.';                              │
│00012        text2='It was overdrawn by '                        │
│                                                              R  │
└─────────────────────────────────────────────────────────────────┘
```

```
┌REPORT──────────────────────────────────────────────────────────┐
│File Edit Search View Locals Globals                             │
│                                                              1  │
│         ┌BREAK─────────────────────────────┐                    │
│         │        At bottom of report       │                    │
│   Qua│ Options                     Color   │  BUDGET      ACTUAL │
│         │   X OVERLINE                BLUE   │  ,000.00  $222,364.71│
│         │   _ DOUBLE OVERLINE    (=) │RED    │  ,700.00   $4,874.58│
┌COMPUTE ─────────────────────────────────────────────────────────┐
│File View Globals                                                │
│                                                                 │
│00013            || trim(left(put(abs(minbal),dollar11.2))) || '.';│
│00014    end;                                                    │
│00015 line ' ';                                                  │
│00016 line 75*'=';                                               │
│00017 line text1 $75.;                                           │
│00018 line text2 $75.;                                           │
│00019 line 75*'=';                                               │
│00020                                                            │
│00021                                                            │
│00022                                                            │
│00023                                                            │
│00024                                                            │
│                                                              R  │
└─────────────────────────────────────────────────────────────────┘
```

You can use a WHERE command to test this report definition for the case
when no departments are over budget. The WHERE command enables you to
select a subset of your data. Here, you want to select departments that are not
over budget so that you can verify that the conditional logic is working.

Select **Search** from the action bar at the top of the REPORT window. From
the pull-down menu that appears, select **Where.** In the Where dialog box enter
the WHERE clause that selects the facilities and travel departments:

```
dept='Facilities' or dept='Travel'
```

When your display looks like Display 6.15, exit the Where dialog box.

Display 6.15
*Selecting
Observations with
the WHERE
Command*

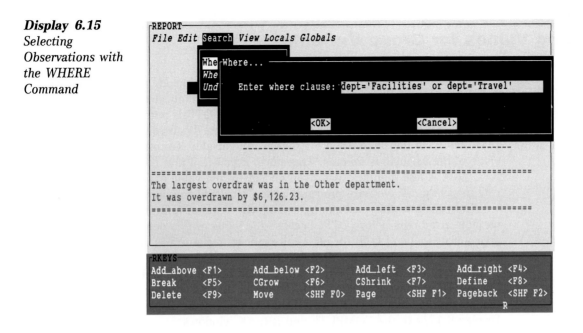

Note: In the Where dialog box you do not specify either the keyword WHERE or a semicolon. If you use the WHERE statement from the command line, you must use the keyword but not the semicolon.

As Display 6.16 illustrates, when no departments are over budget the appropriate text appears in your customized break line.

Display 6.16
*Customized Break
with No
Departments Over
Budget*

```
┌REPORT─────────────────────────────────────────────────────┐
│ File Edit Search View Locals Globals                       │
│                                                          1 │
│                                                            │
│         Quarter  Department     Balance     BUDGET     ACTUAL │
│            1   Facilities    $1,310.16  $31,750.00  $30,439.84 │
│                Travel          $717.59   $4,300.00   $3,582.41 │
│            2   Facilities       $62.71  $30,250.00  $30,187.29 │
│                Travel          $825.42   $5,700.00   $4,874.58 │
│                            ----------  ----------  ---------- │
│                Total:       $2,915.88  $72,000.00  $69,084.12 │
│                            ----------  ----------  ---------- │
│                                                            │
│ ==========================================================  │
│ No departments were over budget.                           │
│                                                            │
│ ==========================================================  │
└────────────────────────────────────────────────────────────┘
┌RKEYS───────────────────────────────────────────────────────┐
│ Add_above <F1>      Add_below <F2>     Add_left  <F3>    Add_right <F4>  │
│ Break     <F5>      CGrow     <F6>     CShrink   <F7>    Define    <F8>  │
│ Delete    <F9>      Move      <SHF F0> Page      <SHF F1> Pageback <SHF F2> │
│                                                          R │
└────────────────────────────────────────────────────────────┘
```

Now cancel the WHERE clause so that you are once again working with the full data set. To cancel the WHERE clause, select **Search.** From the pull-down menu that appears, select **Undo last where.** For further information on the use of the WHERE statement, see Chapter 9, "SAS Language Statements," in *SAS Language: Reference.*

Including Values for Group Variables

Notice that the customized break lines in Output 6.2 include the quarter in which the maximum overdraw occurred. It is a bit more complicated to include this value than it might appear at first glance because QTR is a group variable. The value for a group variable appears only on the first detail row of the set of detail rows that have the same value for that group variable. On other detail rows, the value of the variable is missing. Therefore, if you find that the current value of BALANCE is less than MINBAL, you do not have a value of QTR on that detail row to save unless you are processing the first detail row in a group.

You can solve this problem by storing the value of QTR in a DATA step variable at a break before QTR. Because the value of this new variable is the appropriate value of QTR for all detail rows in a group, you can save it when you save the values of DEPT and BALANCE.

To store the value of QTR, create a break before QTR and open the COMPUTE window associated with the break. In this window enter the statement that stores the value of QTR in HOLDQTR. When your display looks like Display 6.17, exit the COMPUTE and BREAK windows.

Display 6.17
Storing the Value of QTR

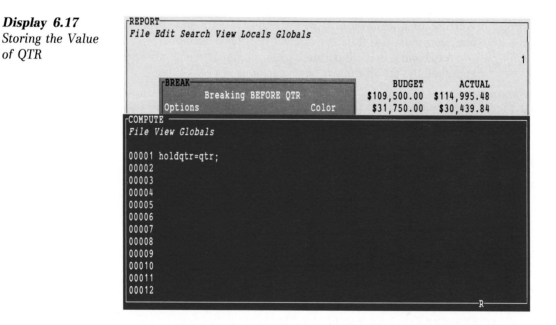

Next, open the DEFINITION window for BALANCE and select the **Edit Program** push button to open the COMPUTE window associated with BALANCE. This window appears in Display 6.18.

Display 6.18
COMPUTE
Window
Associated with
BALANCE

```
┌REPORT──────────────────────────────────────────────────┐
│ File Edit Search View Locals Globals                    │
│                                                        1│
│        Quar┌DEFINITION──────────────────────────────────┐
│            │           Definition of BALANCE            │
│            │   Usage    Attributes        Options    Color│
┌COMPUTE─────────────────────────────────────────────────┐
│ File View Globals                                       │
│                                                         │
│ 00001 balance=budget.sum-actual.sum;                    │
│ 00002 if dept ne ' ' then do;                           │
│ 00003 if balance<minbal then                            │
│ 00004    do;                                            │
│ 00005       minbal=balance;                             │
│ 00006       mindept=dept;                               │
│ 00007    end;                                           │
│ 00008 end;                                              │
│ 00009                                                   │
│ 00010                                                   │
│ 00011                                                   │
│ 00012                                                   │
│                                                       R─│
└─────────────────────────────────────────────────────────┘
```

Add the following statement inside the DO group that assigns values to MINBAL and MINDEPT:

```
minqtr=holdqtr;
```

This statement stores the appropriate value for QTR when you store the values of BALANCE and DEPT. When your display looks like Display 6.19, exit the COMPUTE and DEFINITION windows.

Display 6.19
Storing the
Appropriate Value
of QTR

```
┌REPORT──────────────────────────────────────────────────┐
│ File Edit Search View Locals Globals                    │
│                                                        1│
│        Quar┌DEFINITION──────────────────────────────────┐
│            │           Definition of BALANCE            │
│            │   Usage    Attributes        Options    Color│
┌COMPUTE─────────────────────────────────────────────────┐
│ File View Globals                                       │
│                                                         │
│ 00001 balance=budget.sum-actual.sum;                    │
│ 00002 if dept ne ' ' then do;                           │
│ 00003 if balance<minbal then                            │
│ 00004    do;                                            │
│ 00005       minbal=balance;                             │
│ 00006       mindept=dept;                               │
│ 00007       minqtr=holdqtr;                             │
│ 00008    end;                                           │
│ 00009 end;                                              │
│ 00010                                                   │
│ 00011                                                   │
│ 00012                                                   │
│                                                       R─│
└─────────────────────────────────────────────────────────┘
```

Of course, you also need to modify the statements in the COMPUTE window attached to the break at the end of the report to include the value of MINQTR in the break line. The following statement is a modified version of the assignment statement used to create text for the customized break line in the previous report.

In this statement, the PUT function converts MINQTR to a character string so that you can concatenate it into the character variable TEXT1. The format used in the conversion is the FORQTR3. format, which you created when you invoked the REPORT procedure for the first time in this chapter. Replace the existing assignment statement for TEXT1 with this one:

```
text1='The largest overdraw was in the ' || trim(mindept)
      || ' department during the ' || put(minqtr,forqtr3.)
      || ' quarter.';
```

The modified break appears in Display 6.20.

*Display 6.20
Including the
Value of a Group
Variable in a
Customized Break
Line*

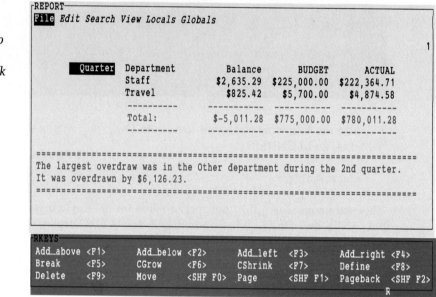

Customizing Break Lines at a Break on a Variable

Now you need to modify the report to determine the department that most overspent its quarterly budget. To do so, you must create two more DATA step variables to hold the values of the minimum balance for each quarter (QMINBAL) and the name of the corresponding department (QMINDEPT). Because you are determining the minimum for each quarter independently, you want to intialize these variables before the first detail row of each group of detail rows.

Initializing Variables

To initialize QMINBAL and QMINDEPT, modify the COMPUTE window attached to the break before QTR to include the statements needed to intialize the new DATA step variables. When your display looks like Display 6.21, exit the COMPUTE and BREAK windows.

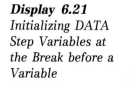

Display 6.21
Initializing DATA Step Variables at the Break before a Variable

```
┌REPORT─────────────────────────────────────────────────────────────────┐
 File Edit Search View Locals Globals

                                                                      1

   ┌BREAK─────────────────────────────────────┐   BUDGET      ACTUAL
   │            Breaking BEFORE QTR            │ $225,000.00 $222,364.71
   │ Options                          Color    │   $5,700.00   $4,874.58
┌COMPUTE ───────────────────────────────────────────────────────────────┐
 File View Globals

 00001 holdqtr=qtr;
 00002 qminbal=0;
 00003 qmindept=' ';
 00004
 00005
 00006
 00007
 00008
 00009
 00010
 00011
 00012
                                                                      R
└────────────────────────────────────────────────────────────────────────┘
```

Note: Although you are using the default values to initialize QMINBAL and QMINDEPT, it is essential that you do the initialization because you want to initialize the variables at the beginning of each group. If you rely on PROC REPORT to perform the default initialization, it initializes them only once.

Determining the Minimum Quarterly Balance

Determining the minimum quarterly balance and the corresponding department is similar to determining the year-to-date values except that you initialize the variables before the break on the break variable instead of at a break at the beginning of the report. Now that you have initialized the variables, you simply add another IF-THEN statement to the COMPUTE window attached to the variable BALANCE. This time the IF-THEN statement compares the value of BALANCE to the value for the minimum quarterly balance, stored in QMINBAL. If BALANCE is less than the current value of QMINBAL, the values of BALANCE and DEPT replace the values in QMINBAL and QMINDEPT.

Add the following statements to the COMPUTE window attached to BALANCE immediately after the END statement that closes the DO group that assigns values to MINBAL, MINDEPT, and MINQTR:

```
if balance<qminbal then
   do;
      qminbal=balance;
      qmindept=dept;
   end;
```

When your display looks like Display 6.22, exit the COMPUTE and DEFINITION windows.

Note: Not all of the code segment is visible in the COMPUTE window at one time. Display 6.22 shows the last part of the code segment, which includes the statements you just added.

Display 6.22
Determining the
Minimum
Quarterly Balance

```
┌REPORT────────────────────────────────────────────────────────────┐
│File Edit Search View Locals Globals                               │
│                                                                   │
│                                                                 1 │
│                                                                   │
│         Quar┌DEFINITION──────────────────────────────────────────┐│
│             │                     Definition of BALANCE           ││
│             │     Usage      Attributes            Options   Color││
│┌COMPUTE ────────────────────────────────────────────────────────┐│
││File View Globals                                                 │
││                                                                  │
││00002 if dept ne ' ' then do;                                     │
││00003 if balance<minbal then                                      │
││00004    do;                                                      │
││00005       minbal=balance;                                       │
││00006       mindept=dept;                                         │
││00007       minqtr=holdqtr;                                       │
││00008    end;                                                     │
││00009 if balance<qminbal then                                     │
││00010    do;                                                      │
││00011       qminbal=balance;                                      │
││00012       qmindept=dept;                                        │
││00013    end;                                                     │
││                                                              R──┘│
└───────────────────────────────────────────────────────────────────┘
```

Writing the Break Lines

To complete this report, you need to write the customized break lines that identify the minimum quarterly balance and the corresponding department. The code segment that does so is comparable to the code segment you used earlier to write the customized break lines at the end of the report. Because no new concepts are involved, the code segment is not explained in detail here.

Select **QTR** and issue the BREAK command. Select **After detail** from the LOCATION window and a color of your choice from the BREAK window. Then select the **Edit Program** push button and put the following SAS statements in the COMPUTE window. The REPORT procedure executes the code segment at the end of each group of rows that have the same value of QTR.

```
length text3 text4 $ 75;
if qminbal=0 then
   do;
      text3='No departments were over budget this quarter.';
      text4=' ';
   end;
else
   do;
      text3='The largest overdraw for this quarter was in the '
            || trim(qmindept) || ' department.';
      text4='It was overdrawn by ' ||
            trim(left(put(abs(qminbal),dollar11.2))) || '.';
   end;
line ' ';
line text3 $75.;
line text4 $75.;
line ' ';
```

When you have finished entering this code segment, exit the COMPUTE and BREAK windows. The modified report appears in Display 6.23.

Display 6.23
Report with Break
Lines after Quarter

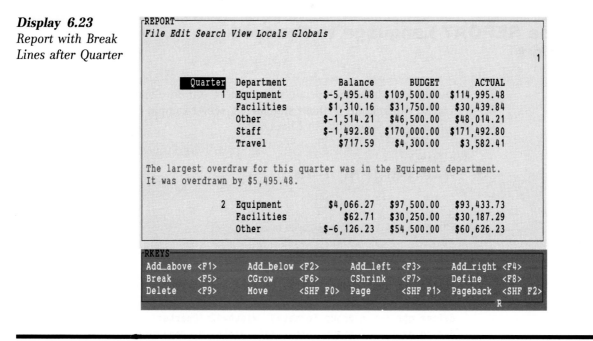

```
┌REPORT─────────────────────────────────────────────────────────────┐
│ File Edit Search View Locals Globals                               │
│                                                                  1 │
│                                                                    │
│         Quarter  Department        Balance        BUDGET      ACTUAL│
│             1    Equipment     $-5,495.48   $109,500.00 $114,995.48│
│                  Facilities     $1,310.16    $31,750.00  $30,439.84│
│                  Other         $-1,514.21    $46,500.00  $48,014.21│
│                  Staff         $-1,492.80   $170,000.00 $171,492.80│
│                  Travel           $717.59     $4,300.00   $3,582.41│
│                                                                    │
│ The largest overdraw for this quarter was in the Equipment department.│
│ It was overdrawn by $5,495.48.                                     │
│                                                                    │
│             2    Equipment      $4,066.27    $97,500.00  $93,433.73│
│                  Facilities        $62.71    $30,250.00  $30,187.29│
│                  Other         $-6,126.23    $54,500.00  $60,626.23│
│                                                                    │
├RKEYS───────────────────────────────────────────────────────────────┤
│ Add_above <F1>      Add_below <F2>      Add_left  <F3>   Add_right <F4>│
│ Break     <F5>      CGrow     <F6>      CShrink   <F7>   Define   <F8>│
│ Delete    <F9>      Move    <SHF F0>    Page    <SHF F1> Pageback <SHF F2>│
│                                                              R      │
└────────────────────────────────────────────────────────────────────┘
```

Adding the Finishing Touches

Type over the column headers to change **BUDGET** to **Budget** and **ACTUAL** to **Actual**. Select **Locals** from the action bar at the top of the REPORT window. Open the ROPTIONS window and select **HEADSKIP.** Then exit the ROPTIONS window. The final report appears in Display 6.24.

Display 6.24
Final Report

```
┌REPORT─────────────────────────────────────────────────────────────┐
│ File Edit Search View Locals Globals                               │
│                                                                  1 │
│                                                                    │
│         Quarter  Department        Balance        Budget      Actual│
│                                                                    │
│             1    Equipment     $-5,495.48   $109,500.00 $114,995.48│
│                  Facilities     $1,310.16    $31,750.00  $30,439.84│
│                  Other         $-1,514.21    $46,500.00  $48,014.21│
│                  Staff         $-1,492.80   $170,000.00 $171,492.80│
│                  Travel           $717.59     $4,300.00   $3,582.41│
│                                                                    │
│ The largest overdraw for this quarter was in the Equipment department.│
│ It was overdrawn by $5,495.48.                                     │
│                                                                    │
│             2    Equipment      $4,066.27    $97,500.00  $93,433.73│
│                  Facilities        $62.71    $30,250.00  $30,187.29│
│                                                                    │
├RKEYS───────────────────────────────────────────────────────────────┤
│ Add_above <F1>      Add_below <F2>      Add_left  <F3>   Add_right <F4>│
│ Break     <F5>      CGrow     <F6>      CShrink   <F7>   Define   <F8>│
│ Delete    <F9>      Move    <SHF F0>    Page    <SHF F1> Pageback <SHF F2>│
│                                                              R      │
└────────────────────────────────────────────────────────────────────┘
```

Use the RSTORE command to store the report definition in
SASUSER.TUTORS.CHAP6.REPT.

Using the REPORT Language to Produce the Report

The following SAS statements produce the report you developed in this chapter.
For a detailed explanation of these statements, see "Example 6: Creating a Report
with Multiple Code Segments" in Chapter 10.

```
proc format;
   value forqtr 1='1st' 2='2nd';
run;

title;

proc report data=report.budget;

   column qtr dept budget actual balance budget=bud2 actual=act2;

   define qtr     / group format=3. width=12 'Quarter';
   define dept    / group format=$10. width=10 'Department';
   define budget  / sum noprint;
   define actual  / sum noprint;
   define balance / computed format=dollar11.2 width=11 'Balance';
   define bud2    / sum format=dollar11.2 width=11 'Budget';
   define act2    / sum format=dollar11.2 width=11 'Actual';

   compute balance;
      balance=budget.sum-actual.sum;
      if dept ne ' ' then
         do;
            if balance<minbal then
               do;
                  minbal=balance;
                  mindept=dept;
                  minqtr=holdqtr;
               end;
            if balance<qminbal then
               do;
                  qminbal=balance;
                  qmindept=dept;
               end;
         end;
   endcomp;

   compute before qtr;
      holdqtr=qtr;
      qminbal=0;
      qmindept=' ';
   endcomp;
```

```
            break after qtr / color=red;

            compute after qtr;
               length text3 text4 $ 75;
               if qminbal=0 then
                  do;
                     text3='No departments were over budget this quarter.';
                     text4=' ';
                  end;
               else
                  do;
                     text3='The largest overdraw for this quarter was in the '
                             || trim(qmindept) || ' department.';
                     text4='It was overdrawn by '
                             || trim(left(put(abs(qminbal),dollar11.2))) || '.';
                  end;
               line ' ';
               line text3 $75.;
               line text4 $75.;
               line ' ';
            endcomp;

            compute before ;
               minbal=0;
               mindept=' ';
            endcomp;

            rbreak after / ol ul summarize color=red;

            compute after;
               dept='Total:';
               length text1 text2 $ 75;
               if minbal=0 then
                  do;
                     text1='No departments were over budget.';
                     text2=' ';
                  end;
               else
                  do;
                     text1='The largest overdraw was in the ' || trim(mindept)
                             || ' department during the '
                             || put(minqtr,forqtr3.) || ' quarter.';
                     text2='It was overdrawn by '
                             || trim(left(put(abs(minbal),dollar11.2))) || '.';
                  end;
               line ' ';
               line 75*'=';
               line text1 $75.;
               line text2 $75.;
               line 75*'=';
            endcomp;

         run;
```

Part 2
Reference

Chapter 7 Using the REPORT Procedure in a Windowing Environment

Introduction

You can use the REPORT procedure in either a windowing or a nonwindowing environment. This chapter and much of this book deal with using PROC REPORT in a windowing environment. Chapter 10, "The REPORT Language," explains how to use the REPORT language, which enables you to use the REPORT procedure without windows.

This chapter explains how to

- [ ] invoke the REPORT procedure
- [ ] use SAS statements with the REPORT procedure
- [ ] select items in the body of the report
- [ ] use windows
- [ ] use action bars and pull-down menus
- [ ] issue commands
- [ ] use the help facility for the REPORT procedure.

Invoking the REPORT Procedure

You invoke the REPORT procedure by submitting the PROC REPORT statement. The syntax for invoking PROC REPORT in a windowing environment follows:

PROC REPORT WINDOWS | PROMPT <*option-list*>;

where

WINDOWS | WD
>   specifies that you want to use the windowing environment. By default, the procedure runs in the nonwindowing environment. When you use the WINDOWS option, the SAS System opens the REPORT window, which enables you to modify a report repeatedly and to see the modifications immediately.

PROMPT
>   opens the REPORT window and starts the PROMPT facility. This facility guides you through creating a new report or adding more data set variables or statistics to an existing report.

The list of options can include the options described below. This list includes only those options most useful in a windowing environment. Chapter 10 describes all options.

CENTER | NOCENTER
>   specifies whether or not to center the report in the REPORT window. You can control the centering of a report from the PROC REPORT statement, from a stored report definition, or from the setting of the CENTER system option. The REPORT procedure first honors the setting in the PROC REPORT statement. If you do not specify how to center the report there, PROC REPORT honors the setting of the CENTER option in a report definition loaded with the REPORT= option in the PROC REPORT statement. If you use neither a procedure statement option nor a stored report definition, PROC REPORT honors the setting of the CENTER system option. You can change centering during a PROC REPORT session with the CENTER option in the ROPTIONS window.
>
>   **Note:**   Loading a report definition during a PROC REPORT session causes the value of the CENTER option stored with the report definition to take effect.

COLWIDTH=*column-width*
>   specifies the default number of horizontal positions for all columns in the report. Valid values range from 1 to the line size. The default value is 9.
>
>   When setting the width for a column, PROC REPORT first looks for a width specification in the definition for that item. If it finds none and the item is a data set variable, it uses a column width large enough to accommodate the format (if one exists) associated with the variable in the input data set. Otherwise, it uses the value of the COLWIDTH= option.

COMMAND
>   displays command lines rather than action bars in all REPORT windows when you invoke the REPORT procedure. After you have invoked the REPORT procedure, you can display the action bars by issuing the COMMAND command.

DATA=*SAS-data-set*

names the SAS data set to use for input. If you do not specify the DATA= option, PROC REPORT uses the data set whose name is stored in the automatic variable _LAST_. This data set is usually the data set most recently created within your SAS session, although you can set the value of _LAST_ yourself with the _LAST_= system option.

HEADLINE

underlines all column headers and the spaces between them at the top of each page of the report.

HEADSKIP

writes a blank line beneath all column headers (or beneath the underlining that the HEADLINE option writes) at the top of each page of the report.

HELP=*libref.catalog*

names the library and catalog containing the user-defined CBT or HELP entries for the report. PROC REPORT stores all HELP entries for a report in the same catalog. You can write a CBT or HELP entry for each item in the report. Specify the entry name from the DEFINITION window for the variable. You create the entry with the BUILD procedure in SAS/AF software. For information on PROC BUILD, see Chapter 10, "The BUILD Procedure," in *SAS/AF Software: Usage and Reference, Version 6, First Edition.*

When you use the HELP= option, you must specify the following:

libref   points to the SAS data library containing the HELP catalog. You can use any valid libref in this field, but be sure to define the libref before invoking the REPORT procedure. Use the LIBNAME statement to associate a libref with the SAS data library. For details on the LIBNAME statement, see *SAS Language: Reference, Version 6, First Edition.*

catalog   points to a catalog in *libref.*

LS=*line-size*

specifies the length (the number of horizontal positions) of a line of the report. Valid values range from 64 to 256. You can control the line size of a report from the PROC REPORT statement, from a stored report definition, or from the LINESIZE= system option. The REPORT procedure first honors the setting in the PROC REPORT statement. If you do not specify a line size for the report there, PROC REPORT honors the setting of the Linesize attribute in a report definition loaded with the REPORT= option in the PROC REPORT statement. If you use neither a procedure statement option nor a stored report definition, PROC REPORT honors the setting of the LINESIZE= system option. You can change the line size during a PROC REPORT session with the Linesize attribute in the ROPTIONS window.

Note: Loading a report definition during a PROC REPORT session causes the value of the Linesize attribute stored with the report definition to take effect.

MISSING

considers missing values as valid values for group, order, or across variables. Special missing values used to represent numeric values (the letters A through Z and the underscore) are each considered as a different value. A group for each missing value appears in the report. If you do not specify the MISSING

(MISSING continued)

option, PROC REPORT does not include observations with a missing value for one or more group, order, or across variables in the report.

NORKEYS

suppresses the initial display of the RKEYS window. After you have invoked the REPORT procedure, you can use the RKEYS command to display the RKEYS window.

OUTREPT=*libref.catalog.entry*

names the entry in which to store the report definition in use when the REPORT procedure terminates.

When you use the OUTREPT= option, you must specify the following:

libref   points to a SAS data library. Use the LIBNAME statement to associate a libref with the SAS data library. For details on the LIBNAME statement, see *SAS Language: Reference.*

catalog   names a SAS catalog in *libref.*

entry   names the entry to which to write the report definition.

The SAS System assigns an entry type of REPT to the entry.

Note: You can use the RSTORE command to save report definitions from the REPORT window before you are ready to terminate the procedure.

PANELS=*number-of-panels*

specifies the number of panels on each page of the report. If the width of a report is less than half of the line size, you can display the data in multiple sets of columns so that rows that would otherwise appear on multiple pages appear on the same page. Each set of columns is called a *panel.* A familiar example of this kind of report is a telephone book, which contains multiple panels of names and telephone numbers on a single page.

By default, the REPORT procedure creates a report with one panel per page. For an example of a report that uses multiple panels, see the Panels attribute in the documentation for the ROPTIONS window.

PROFILE=*libref.catalog*

locates your REPORT profile. A profile enables you to customize certain aspects of the report procedure. For information on using a REPORT profile, see the documentation for the PROFILE window.

When you use the PROFILE= option, you must specify the following:

libref   points to a SAS data library. You can use any valid libref in this field, but be sure to define the libref before invoking the REPORT procedure. Use the LIBNAME statement to associate a libref with the SAS data library. For details on the LIBNAME statement, see *SAS Language: Reference.*

catalog   names a SAS catalog in *libref.*

When you use the PROFILE= option, PROC REPORT uses the entry REPORT.PROFILE in the catalog you specify as your profile. If no such entry exists, or if you do not specify a profile, PROC REPORT uses the profile in

SASUSER.PROFILE. If you have no profile, PROC REPORT uses the defaults for all options, colors, pull-down menus, and action bars.

Note: When you create a profile, PROC REPORT stores it in SASUSER.PROFILE.REPORT.PROFILE. Use the CATALOG procedure or the CATALOG window if you want to copy the profile to another location.

PS=*page-size*

specifies the number of lines in a page of the report. Valid values range from 15 to 32,767. You can control the page size of a report from the PROC REPORT statement, from a stored report definition, or from the setting of the PAGESIZE= system option. The REPORT procedure first honors the setting in the PROC REPORT statement. If you do not specify a page size for the report there, PROC REPORT honors the setting of the Pagesize attribute in a report definition loaded with the REPORT= option in the PROC REPORT statement. If you use neither a procedure statement option nor a stored report definition, PROC REPORT honors the PAGESIZE= system option. You can change the page size during a PROC REPORT session with the Pagesize attribute in the ROPTIONS window.

Note: Loading a report definition during a PROC REPORT session causes the value of the Pagesize attribute stored with the report definition to take effect.

PSPACE=*space-between-panels*

specifies the number of blank characters between panels. The default value is 4.

The REPORT procedure separates all panels in the report by the same number of blank characters. For each panel, the sum of the width of the panel and the number of blank characters separating it from the panel to its left cannot exceed the line size.

For information on altering the amount of space between panels after you have invoked the REPORT procedure, see the Pspace attribute in the documentation for the ROPTIONS window.

REPORT=*libref.catalog.entry*

names the report definition to load initially. The REPORT procedure stores all report definitions as entries of type REPT in a SAS catalog. When you use the REPORT= option, you must specify the following:

libref	points to a SAS data library. Use the LIBNAME statement to associate a libref with the SAS data library. For details on the LIBNAME statement, see *SAS Language: Reference*.
catalog	names a SAS catalog in *libref*.
entry	names the entry that contains the report definition you want.

If you don't use the REPORT= option and the input data set contains both character and numeric variables, the REPORT procedure creates a report similar to the output generated by a simple PROC PRINT step. If you don't use the REPORT= option and the input data set contains only numeric variables, PROC REPORT produces a one-line summary report that shows the sum of each variable over all observations in the data set. However, in either case, in the windowing environment, PROC REPORT initially uses only as many of the variables in the data set as can fit on one page of the report. You can easily add other variables to the report.

(REPORT= continued)

For information on loading a different report after you have invoked the REPORT procedure, see the documentation for the RLOAD command.

SHOWALL
overrides selections in a report definition (specified in the REPORT= option) that suppress the display of a column. See the NOPRINT and NOZERO options in the documentation for the DEFINITION window.

SPACING=*space-between-columns*
specifies the number of blank characters between columns. The default value is 2.

The REPORT procedure separates all columns in the report by the number of blank characters specified in the SPACING= option in the PROC REPORT statement unless you use the Spacing attribute in the DEFINITION window to change the spacing to the left of a specific item. For each column, the sum of the column width and number of blank characters separating that column from the column to its left cannot exceed the line size.

For information on altering the amount of space used between columns after you have invoked the REPORT procedure, see the Spacing attribute in the documentation for the ROPTIONS window.

SPLIT=*'character'*
specifies the split character. If you use the split character in a header, the REPORT procedure breaks the header when it reaches that character and continues the header on the next line. The split character itself is not part of the header.

By default, PROC REPORT uses the slash (/) as the split character.

Using SAS Statements with the REPORT Procedure

You can use with PROC REPORT any of the SAS statements that you can normally use with a procedure (see Appendix 1, "SAS Statements Used with Procedures," in the *SAS Procedures Guide, Version 6, Third Edition*). In addition, you can use the FREQ and WEIGHT statements. For details on using these statements with the REPORT procedure, see Chapter 10.

Note: Although you can use the BY statement with the REPORT procedure in a nonwindowing environment, you cannot use the BY statement in a windowing environment.

Note: You can also use statements from the REPORT language (see Chapter 10) to define the initial report in a session. Or, you can use these statements in conjunction with the REPORT= option to modify an existing definition.

Selecting Items in a Report

Often when you modify a report, you select an item in the report and issue a REPORT command. For instance, if you want to redefine the characteristics of a variable in the report, you first select the variable, then issue the DEFINE command.

You select an item by moving the cursor to the desired location and pressing ENTER, RETURN, or the equivalent mouse button. When you select an item, the REPORT procedure displays it in reverse video, that is, it uses the foreground color for the background and the background color for the foreground. To cancel a selection, move the cursor to the highlighted area and press ENTER, RETURN, or the equivalent mouse button.

An *item* is a data set variable, a statistic, or a computed variable. In many cases an item occupies a single column of the report. However, as you can see in Display 7.1, an item can occupy multiple columns. Here, the item **DEPT** is an across variable occupying five columns in the report, one column for each value of the variable DEPT (`Equipment`, `Facilities`, `Other`, `Staff`, and `Travel`). When you select **DEPT** or one of the values of DEPT, the REPORT procedure highlights the column headers for all five columns.

Display 7.1
Items Occupying
Multiple Columns

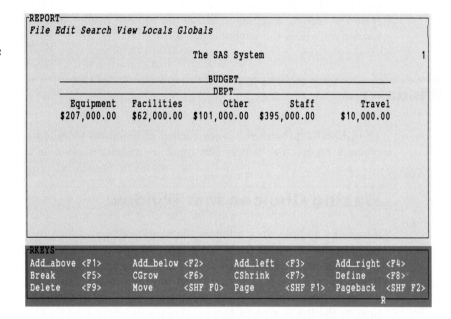

The variable BUDGET, which appears above DEPT in the report, is also a single item. It shares multiple columns with DEPT.

How you choose to move your cursor depends on the equipment available to you. The list below describes some of the more common ways of moving the cursor.

cursor keys	move the cursor in the direction indicated on the key.
a mouse	moves the cursor in the direction in which you move the mouse.
TAB key	moves the cursor to the next radio button, check box, or field in a window.
NEXTFIELD command	moves the cursor to the next column header in the REPORT window or to the next radio button, check box, or field in other windows.

PREVFIELD command	moves the cursor to the previous column header in the REPORT window or to the previous radio button, check box, or field in other windows.
HOME key	moves the cursor to the action bar or command line of a window.

You can select an item in the report by selecting the text in the column header or by selecting a value in the column. If you select the text in the column header, PROC REPORT selects only the column header (or headers if the item appears in more than one column). If you select a value in the column, PROC REPORT selects the entire column (or columns). In most cases it makes no difference how you select an item.

▶ *Caution* . . . . *Selecting an Item for Deletion*
When you use the DELETE command on an item with a column header that contains more than one line of text, selecting the entire column deletes the item from the report, whereas selecting the column header deletes only the line of the header on which the cursor sits. ▲

Using Windows

The REPORT procedure uses a variety of windows to solicit your input as you create a report. You supply this input by making choices or by typing in values.

Making Choices in a Window

You make a choice in a window the same way you select an item in the report: position the cursor at the desired location and press ENTER, RETURN, or the equivalent mouse button. In some cases you can select only one item from a list of items; in other cases, you can select multiple items from a list.

Lists from which you can select only one item are called *radio boxes*. Each item in the list is a *radio button*. When you select a radio button, the REPORT procedure fills in the open circle to its left.

Lists from which you can select more than one item appear in one of two forms:

□ as a list of choices, each of which is preceded by a small box. Each choice in this kind of list is a *check box*.

□ as a simple list of choices.

When you select a check box, the box preceding the choice contains an X, a check mark (✔), or some other character. When you make choices from a simple list, PROC REPORT highlights your choices. If you are adding a data set variable or a statistic to a report, your first choice moves to the top of the list, your second choice to the second row of the list, and so forth.

Supplying Text in a Window

In some cases a window enables you to specify a value for some characteristic of the report, such as the line size. You do so by entering text in the appropriate field.

Scrolling in a Window

Windows such as the DATASET VARS window and the RLOAD window may contain more information than appears on your display. To scroll through the contents of such a window, use the SAS global commands FORWARD and BACKWARD. These commands are in the pull-down menu associated with **View** in the action bar at the top of any window that has an action bar.

Exiting a Window

When you finish using a window, you exit it. You always have at least two commands to choose from when you exit a window: the OK command (also called the END command) and the CANCEL command. The OK command implements the choices you make and the values you specify while you are in the window. Typically, the CANCEL command cancels any changes you have made in the window since you opened it. In the PROMPTER window, however, the CANCEL command cancels prompting at that point.

The OK and CANCEL commands are located in one of two places:

□ in the pull-down menu associated with **File** in the action bar at the top of the window. (For information on action bars, see "Using Action Bars and Pull-Down Menus" later in this chapter.)

□ in push buttons at the bottom of the window. A *push button* is a part of a window that is always highlighted. Selecting a push button initiates an action, such as exiting a window.

Some windows support additional ways of exiting. These ways are documented with the windows that use them.

Note: Pressing a function key defined as the END command is equivalent to selecting the OK push button; pressing a function key defined as the CANCEL command is equivalent to selecting the CANCEL push button. For information on defining function keys, see "Defining a Function Key" later in this chapter.

Summary of Windows

The following list briefly describes what each window in PROC REPORT does:

ADDING
>   displays the kinds of items you can add to a report.

BREAK
>   presents choices controlling the REPORT procedure's actions at breaks (changes) in the values of an order or group variable or at the beginning or end of the report.

COMPUTE
>   uses a subset of DATA step statements and the LINE statement to define computed variables, perform calculations, and customize break lines.

COMPUTED VAR
>   prompts you for the name of the computed variable you are adding to the report.

DATASET VARS
>   lists the variables in the data set and enables you to select the ones you want to add to the report.

DEFINITION
>   displays the characteristics associated with an item in the report and enables you to change them.

FORMAT
>   displays several commonly used SAS formats for you to select from.

LOCATION
>   displays the choices for the location of break lines.

MESSAGE
>   displays notes, warnings, and errors issued by the REPORT procedure, as well as the REPORT language generated by the LIST command.

PROFILE
>   enables you to customize certain aspects of the REPORT procedure.

PROMPTER
>   prompts you for information as you build a report.

REPORT
>   contains the text of the report.

RKEYS
>   displays definitions for up to 12 function keys defined as REPORT commands.

RLOAD
>   displays the names of all report entries stored in the specified catalog and enables you to load the report you select.

ROPTIONS
>   controls layout and display of the entire report and identifies the SAS data library and catalog containing CBT or HELP entries for items in the report.

RSTORE
> prompts you for the complete name of the catalog entry in which to store the definition of the current report.

STATISTICS
> displays the choices for adding statistics to the report.

Using Action Bars and Pull-Down Menus

An *action bar* is a list of selections across the top of a window. The action bars used in REPORT windows contain one or more of the following selections: File, Edit, Search, View, Locals, and Globals. When you make a selection from the action bar, the REPORT procedure displays a pull-down menu that lists REPORT commands, SAS global commands frequently used in PROC REPORT, or both. Select the command you want to issue from the pull-down menu.

Note: The next section, "Issuing Commands from the REPORT Procedure," explains how to issue a SAS global command that does not appear in a pull-down menu.

If you make the wrong selection from an action bar, simply move the cursor out of the pull-down menu that appears and press ENTER, RETURN, or the appropriate mouse button. The pull-down menu disappears.

You can turn off the display of the action bar in the REPORT window without affecting the display in any other window by invoking the COMMAND command, which is located under **Globals** in the action bar. To turn off the display of action bars in all windows, first turn off their display in the REPORT window, then issue the PMENU command. To redisplay an action bar, enter the same command (COMMAND or PMENU) on the command line.

Issuing Commands from the REPORT Procedure

You can issue a command from the REPORT procedure in any of the following ways:

□ by pressing a function key. The REPORT procedure provides a set of default key definitions that you can alter through the KEYS window. (To open the KEYS window, issue the KEYS command, which is in the pull-down menu under **Globals** in the action bar at the top of any window with an action bar.) The RKEYS window, which appears below the REPORT window by default, displays definitions for up to 12 function keys defined as REPORT commands.

For information on changing the default function key definitions, see "Defining a Function Key" later in this chapter.

□ by making a selection from an action bar at the top of the REPORT window and, when the corresponding pull-down menu appears, selecting the command you want to issue.

□ by turning off the display of action bars with either the COMMAND or PMENU command, typing the name of the command (and arguments if appropriate) on the command line, and pressing ENTER or RETURN. Some commands that open windows, like the RSTORE command, enable you to bypass the window by specifying arguments when you issue the command. To do so, you must issue the command from the command line or use a function key defined as the command with the appropriate arguments.

□   by selecting the name of the command or the name of the corresponding function key from the RKEYS window.

Summary of Commands

The following list briefly describes what each REPORT command does:

ADD_ABOVE
   adds one or more data set variables or statistics above the selected item; or adds a blank line above the first line of the selected header.

ADD_BELOW
   adds one or more data set variables or statistics below the selected item; or adds a blank line below the last line of the selected header.

ADD_LEFT
   adds one or more data set variables, one or more statistics, or a computed variable to the left of the selected item.

ADD_RIGHT
   adds one or more data set variables, one or more statistics, or a computed variable to the right of the selected item.

BREAK
   controls the REPORT procedure's actions at a break (change) in the value of a group or order variable.

CGROW
   increases by one horizontal position the width of the column or columns containing the selected item.

CSHRINK
   decreases by one horizontal position the width of the column or columns containing the selected item.

DEFINE
   displays the characteristics of the selected item.

DELETE
   removes an item from the report or a line from a header.

LIST
   writes the REPORT language defining the current report to the MESSAGE window.

MESSAGE
   opens the MESSAGE window, which contains notes, warnings, and errors issued by the REPORT procedure, as well as the REPORT language generated by the LIST command.

MOVE
   moves an item from one location in the report to another.

PAGE
: displays the next page of the report.

PAGEBACK
: displays the previous page of the report.

PROFILE
: enables you to customize certain aspects of the REPORT procedure.

QUIT
: terminates the REPORT procedure.

RBREAK
: controls the REPORT procedure's actions at a break at the beginning or end of the report.

REFRESH
: displays the most recent version of the report.

RKEYS
: displays the RKEYS window if it is not currently displayed and refreshes the definitions in the RKEYS window with the most recently created ones.

RLOAD <*libref.catalog.entry*>
: loads the specified report definition or, if you do not specify an argument, displays a selection of report definitions.

ROPTIONS
: opens the ROPTIONS window, which controls the layout and display of the entire report and identifies the SAS data library and catalog containing CBT or HELP entries for items in the report.

RSTORE <*libref.catalog.entry*>
: stores a report definition in the specified catalog entry or, if you do not specify an argument, opens the RSTORE window, which prompts you for a catalog entry.

SPAN
: produces a header that spans multiple columns.

Using the Help Facility

The REPORT procedure provides an extensive, context-sensitive help facility. You can use the help facility to get information about

□ the general nature of the REPORT procedure.

□ specific features of the REPORT procedure, such as individual options and windows.

□ specific items in the report. This kind of help is a user-defined HELP or CBT entry. You create the entries with the BUILD procedure in SAS/AF software and point to them in the appropriate fields in the ROPTIONS and DEFINITION windows for the item. For information on PROC BUILD, see Chapter 10 in *SAS/AF Software: Usage and Reference*.

The text in CBT entries may include words and groups of words that are highlighted in some way. The exact method of highlighting varies from one operating system to another. Typically, the text is in a different color from most of the text or in reverse video. You can get help on these topics by moving the cursor to the highlighted text you are interested in and pressing the HELP function key, or ENTER, RETURN, or the corresponding mouse button.

Note: By default, PROC REPORT defines a HELP function key for you. To determine which key is the HELP function key, issue the KEYS command, which is in the pull-down menu under **Globals** in the action bar at the top of any window with an action bar.

To get help with the REPORT procedure in general, issue the HELP command or press the HELP function key when the cursor is not on a specific item in the report. You can get help with individual report commands by selecting **Report commands** from the main REPORT help frame.

Sometimes you may want to get help with a specific window or with a field in a window. In these cases, you must use the HELP function key.

For instance, to get help on **Usage** in the DEFINITION window, move the cursor to the word **Usage** and press the HELP function key.

To get help with an item in the report (a data set variable, a statistic, or a computed variable), position the cursor on the item and press the HELP function key. Because HELP entries for items in the report are user-defined, the report may contain items for which no HELP entry exists. If you request help with such an item, the REPORT procedure displays the HELP entry for the REPORT window.

Defining a Function Key

The REPORT procedure provides a set of default key definitions. The names of the first 12 function keys defined as REPORT commands appear in the RKEYS window. You can alter these definitions or define additional function keys through the KEYS window.

You can assign any REPORT command or any SAS global command to a function key. It is particularly useful to assign the END and CANCEL commands to function keys so that you can exit a window with one keystroke. If you don't have a mouse, it is convenient to assign the FORWARD and BACKWARD commands to function keys. To define a function key, proceed as follows:

1. Select **Globals** from the action bar at the top of any window with an action bar.

2. Select **KEYS** from the pull-down menu that appears.

3. The REPORT procedure displays the KEYS window, which appears in Display 7.2. The KEYS window lists the functions keys available to you and their definitions within the REPORT procedure.
   Note: The definitions in the KEYS window are system-dependent.
   Enter the name of a command and any arguments in the Definition field next to the name of the appropriate function key.

4. Select **File** from the action bar at the top of the KEYS window, and select **END** from the pull-down menu.

Display 7.2
Displaying the
Current Function
Key Definitions in
the KEYS Window

```
┌REPORT──────────────────────────────┐ ┌KEYS <REPORT>──────────────────┐
│File Edit Search View Locals Globals│ │File Edit View Globals Help    │
│                                    │ │                               │
│                       The SAS      │ │Key        Definition          │
│                                    │ │                               │
│             DEPT      ACCOUN        │ │F0         HELP                │
│             Equipment lease         │ │F1         ADD_ABOVE           │
│                       maint         │ │F2         ADD_BELOW           │
│                       purcha        │ │F3         ADD_LEFT            │
│                       rental        │ │F4         ADD_RIGHT           │
│                       sets          │ │F5         BREAK               │
│                       tape          │ │F6         CGROW               │
│             Facilities rent         │ │F7         CSHRINK             │
│                       suppli        │ │F8         DEFINE              │
│                       utils         │ │F9         DELETE              │
│             Other     advert        │ │SHF F0     MOVE                │
│                       musicf        │ │SHF F1     PAGE                │
│                       talent        │ │SHF F2     PAGEBACK            │
│                                    │ │SHF F3                         │
├RKEYS───────────────────────────────┤ │SHF F4                         │
│Add_above <F1>      Add_below <F2>  │ │SHF F5                         │
│Break     <F5>      CGrow     <F6>  │ │SHF F6                         │
│Delete    <F9>      Move      <SHF F0>│ │SHF F7                       │
└────────────────────────────────────┘ └───────────────────────────R───┘
```

You must issue the RKEYS command to see the new definitions in the RKEYS window (see the documentation on the RKEYS command for details).

Chapter 8 REPORT Commands

ADD_ABOVE

Adds one or more data set variables or statistics above the selected item; or adds a blank line above the first line of the selected header

Action Bar

Edit

Description

The ADD_ABOVE command has three functions:

☐ adding one or more data set variables above the selected item

☐ adding one or more statistics above the selected item

☐ adding a blank line above the first line of a header.

If you add several variables above one another, you must use all or all but one of them as across variables (see "Usage" in the documentation for the DEFINITION window). PROC REPORT displays a column for each value of each across variable you add (see Output 8.2).

ADD_ABOVE *continued*

If you add one or more statistics above a variable, PROC REPORT displays a column for each statistic you add.

Usage

1. Select an item.

2. Issue the ADD_ABOVE command.

3. Select **Dataset variable, Statistic,** or **Header line** from the ADDING window. Depending on your choice, PROC REPORT leads you down one of three paths:

   □ If you select **Dataset variable,** PROC REPORT displays the DATASET VARS window. Select one or more variables from this window. When the DEFINITION window appears on your display, select the characteristics you want for each variable. Remember, **Usage** should probably be **ACROSS.**

   □ If you select **Statistic,** PROC REPORT displays the STATISTICS window. Select one or more statistics from this window.

   □ If you select **Header line,** PROC REPORT adds a blank line above the first line of the selected header and highlights the entire header. Typing text in this highlighted area changes the header. The REPORT procedure justifies the header and honors the split character (defined in the ROPTIONS window) when you issue the REFRESH command or redefine the report by adding or deleting an item, by changing information in the DEFINITION window, or by changing information in the BREAK window.

Illustration

Output 8.1 and Output 8.2 illustrate adding the variable QTR above the variable BUDGET. Output 8.1 shows each department's total for BUDGET for the year to date. DEPT is a group variable, and BUDGET is an analysis variable used to calculate the SUM statistic (see "Usage" in the documentation for the DEFINITION window).

Output 8.1
Report without QTR

```
                        Initial Report                      1

            DEPT            BUDGET
            Equipment   $207,000.00
            Facilities   $62,000.00
            Other       $101,000.00
            Staff       $395,000.00
            Travel       $10,000.00
```

After you add QTR as an across variable above BUDGET, PROC REPORT shows each department's total for BUDGET for each quarter.

Output 8.2
Adding a Variable above BUDGET

```
                 Report after Adding QTR above BUDGET              1

                                  QTR

                             1                    2
         DEPT              BUDGET               BUDGET
         Equipment      $109,500.00          $97,500.00
         Facilities      $31,750.00          $30,250.00
         Other           $46,500.00          $54,500.00
         Staff          $170,000.00         $225,000.00
         Travel           $4,300.00           $5,700.00
```

Output 8.3 and Output 8.4 illustrate adding the statistics MIN and MAX above the variable BUDGET. Output 8.3, which groups the data by quarter and department, shows the total budget for all accounts in each department for each quarter. QTR and DEPT are group variables, and BUDGET is an analysis variable used to calculate the SUM statistic.

Output 8.3
Report without Statistics

```
                 Report without Statistics                        1

            QTR  DEPT            BUDGET
              1  Equipment    $109,500.00
                 Facilities    $31,750.00
                 Other         $46,500.00
                 Staff        $170,000.00
                 Travel         $4,300.00
              2  Equipment     $97,500.00
                 Facilities    $30,250.00
                 Other         $54,500.00
                 Staff        $225,000.00
                 Travel         $5,700.00
```

After you add MIN and MAX above BUDGET, the report shows the minimum and maximum budget figures for each department for each quarter.

Output 8.4
Adding Statistics above BUDGET

```
                 Report after Adding MIN and MAX above BUDGET     1

                                 MIN          MAX
            QTR  DEPT          BUDGET       BUDGET
              1  Equipment    $4,000.00   $40,000.00
                 Facilities   $2,750.00   $24,000.00
                 Other        $3,000.00   $30,000.00
                 Staff       $40,000.00  $130,000.00
                 Travel         $800.00    $3,500.00
              2  Equipment    $6,000.00   $40,000.00
                 Facilities   $2,750.00   $24,000.00
                 Other        $5,000.00   $30,000.00
                 Staff       $60,000.00  $165,000.00
                 Travel       $1,200.00    $4,500.00
```

When you add statistics above or below an analysis variable, the REPORT procedure no longer displays the statistic associated with the analysis variable in the DEFINITION window. In this case, PROC REPORT displays MIN and MAX, the statistics added above BUDGET, but no longer displays SUM, the statistic specified in the DEFINITION window for BUDGET.

In the COMPUTE window, you refer to the value of a statistic with a compound name, whether you added that statistic to the report from the STATISTICS window or associated it with an analysis variable in the

ADD_ABOVE *continued*

DEFINITION window. For example, refer to the statistics in Output 8.4 as BUDGET.MIN and BUDGET.MAX.

Note: You can distinguish statistics added above or below a variable from a statistic associated with the variable in the DEFINITION window by the default header on the item. If you add a statistic above or below a variable, PROC REPORT adds a second header, the name of the statistic, above or below the name of the variable. If you associate the statistic with the variable in the DEFINITION window, the REPORT procedure uses only one header, the name of the variable.

See Also

the following commands:

- □ ADD_BELOW
- □ ADD_LEFT
- □ ADD_RIGHT

the following windows:

- □ ADDING
- □ DATASET VARS
- □ DEFINITION
- □ STATISTICS

ADD_BELOW

Adds one or more data set variables or statistics below the selected item; or adds a blank line below the last line of the selected header

Action Bar

Edit

Description

The ADD_BELOW command has three functions:

- □ adding one or more data set variables below the selected item
- □ adding one or more statistics below the selected item
- □ adding a blank line below the last line of a header.

If you add several variables below one another, you must use all or all but one of them as across variables (see "Usage" in the documentation for the DEFINITION window). PROC REPORT displays a column for each value of each across variable you add (see Output 8.2).

If you add one or more statistics below a variable, PROC REPORT displays a column for each statistic you add.

Usage

1. Select an item.

2. Issue the ADD_BELOW command.

3. Select **Dataset variable, Statistic,** or **Header line** from the ADDING window. Depending on your choice, PROC REPORT leads you down one of three paths:

   □ If you select **Dataset variable,** PROC REPORT displays the DATASET VARS window. Select one or more variables from this window. When the DEFINITION window appears on your display, select the characteristics you want for each variable. Remember, **Usage** should probably be **ACROSS.**

   □ If you select **Statistic,** PROC REPORT displays the STATISTICS window. Select one or more statistics from this window.

   □ If you select **Header line,** PROC REPORT adds a blank line below the last line of the selected header and highlights the entire header. Typing text in this highlighted area changes the header. The REPORT procedure justifies the header and honors the split character (defined in the ROPTIONS window) when you issue the REFRESH command or redefine the report by adding or deleting an item, by changing information in the DEFINITION window, or by changing information in the BREAK window.

Illustration

Output 8.5 and Output 8.6 illustrate adding the variable QTR below the variable BUDGET. Output 8.5 shows each department's total for BUDGET for the year to date. DEPT is a group variable, and BUDGET is an analysis variable used to calculate the SUM statistic (see "Usage" in the documentation for the DEFINITION window).

Output 8.5
Report without
QTR

```
┌─────────────────────────────────────────────────────────────────┐
│                        Initial Report                    1        │
│                                                                   │
│            DEPT           BUDGET                                  │
│            Equipment   $207,000.00                                │
│            Facilities   $62,000.00                                │
│            Other       $101,000.00                                │
│            Staff       $395,000.00                                │
│            Travel       $10,000.00                                │
│                                                                   │
└─────────────────────────────────────────────────────────────────┘
```

After you add QTR as an across variable below BUDGET, PROC REPORT shows each department's total for BUDGET for each quarter.

ADD_BELOW *continued*

Output 8.6
Adding a Variable below BUDGET

```
                     Report after Adding QTR below BUDGET                 1

                                     BUDGET
                                      QTR
          DEPT              1                   2
          Equipment     $109,500.00        $97,500.00
          Facilities     $31,750.00        $30,250.00
          Other          $46,500.00        $54,500.00
          Staff         $170,000.00       $225,000.00
          Travel          $4,300.00         $5,700.00
```

Output 8.7 and Output 8.8 illustrate adding the statistics MIN and MAX below the variable BUDGET. Output 8.7, which groups the data by quarter and department, shows the total budget for all accounts in each department for each quarter. QTR and DEPT are group variables, and BUDGET is an analysis variable used to calculate the SUM statistic.

Output 8.7
Report without Statistics

```
                       Report without Statistics                        1

          QTR  DEPT              BUDGET
           1   Equipment     $109,500.00
               Facilities     $31,750.00
               Other          $46,500.00
               Staff         $170,000.00
               Travel          $4,300.00
           2   Equipment      $97,500.00
               Facilities     $30,250.00
               Other          $54,500.00
               Staff         $225,000.00
               Travel          $5,700.00
```

After you add MIN and MAX below BUDGET, the report shows the minimum and maximum budget figures for each department for each quarter.

Output 8.8
Adding Statistics below BUDGET

```
                Report after Adding MIN and MAX below BUDGET             1

                                     BUDGET
          QTR  DEPT            MIN              MAX
           1   Equipment    $4,000.00       $40,000.00
               Facilities   $2,750.00       $24,000.00
               Other        $3,000.00       $30,000.00
               Staff       $40,000.00      $130,000.00
               Travel         $800.00        $3,500.00
           2   Equipment    $6,000.00       $40,000.00
               Facilities   $2,750.00       $24,000.00
               Other        $5,000.00       $30,000.00
               Staff       $60,000.00      $165,000.00
               Travel       $1,200.00        $4,500.00
```

When you add statistics above or below an analysis variable, the REPORT procedure no longer displays the statistic associated with the analysis variable in the DEFINITION window. In this case, PROC REPORT displays MIN and MAX, the statistics added below BUDGET, but no longer displays SUM, the statistic specified in the DEFINITION window for BUDGET.

In the COMPUTE window, you refer to the value of a statistic with a compound name, whether you added that statistic to the report from the STATISTICS window or associated it with an analysis variable in the DEFINITION window. For example, refer to the statistics in Output 8.8 as BUDGET.MIN and BUDGET.MAX.

Note: You can distinguish statistics added above or below a variable from a statistic associated with the variable in the DEFINITION window by the default header on the item. If you add a statistic above or below a variable, PROC REPORT adds a second header, the name of the statistic, above or below the name of the variable. If you associate the statistic with the variable in the DEFINITION window, the REPORT procedure uses only one header, the name of the variable.

See Also

the following commands:

□ ADD_ABOVE

□ ADD_LEFT

□ ADD_RIGHT

the following windows:

□ ADDING

□ DATASET VARS

□ DEFINITION

□ STATISTICS

ADD_LEFT

Adds one or more data set variables, one or more statistics, or a computed variable to the left of the selected item

Action Bar

Edit

Description

The ADD_LEFT command adds one or more data set variables, one or more statistics, or a computed variable to the left of the selected item.

Usage

1. Select an item.

2. Issue the ADD_LEFT command.

ADD_LEFT *continued*

3. Select **Dataset variable**, **Statistic**, or **Computed variable** from the ADDING window. Depending on your choice, PROC REPORT leads you down one of three paths:

□ If you select **Dataset variable**, PROC REPORT displays the DATASET VARS window, which lists all the variables in the data set. Select one or more variables from this window. When you exit this window, PROC REPORT opens in turn the DEFINITION window for each variable you selected. In each DEFINITION window, select the characteristics you want for that variable.

□ If you select **Statistic**, PROC REPORT displays the STATISTICS window. Select one or more statistics from this window.

In general, when you add a statistic to a report, the column containing the statistic must be associated with a variable; that is, the statistic must either be named in the Statistic field in the DEFINITION window for an analysis variable or appear above or below a display, analysis, or across variable. If you try to add a statistic to the left of a variable, PROC REPORT has no variable to associate with the statistic. The only meaningful selection from the STATISTICS window in such a case is **N**, a simple count of observations, which is independent of any variable. If you select **N**, each value in the new column represents the number of observations described in that row.

If you are adding statistics to the left of another statistic, you can select any statistics. The REPORT procedure associates the new statistics with the variable associated with the selected statistic.

□ If you select **Computed variable**, PROC REPORT displays the COMPUTED VAR window. Enter a name for the computed variable at the prompt. If you are adding a character variable, select the **Character data** check box and, if you wish, specify the length of the variable at the prompt. (The default length is 8.) Then, select the **Edit Program** push button, which opens the COMPUTE window. Use DATA step statements in the COMPUTE window to define the computed variable.

Note: Computed variables exist only in the report. The REPORT procedure does not add them to the data set.

Illustration

Output 8.9 and Output 8.10 illustrate adding the variable QTR to the left of the variable DEPT. DEPT and QTR are both display variables.

Output 8.9
Report with One
Variable

```
                          Initial Report                          1

                          DEPT
                          Staff
                          Staff
                          Staff
                          Staff
                          Equipment
```

Output 8.10
Report after
Adding QTR to the
Left of DEPT

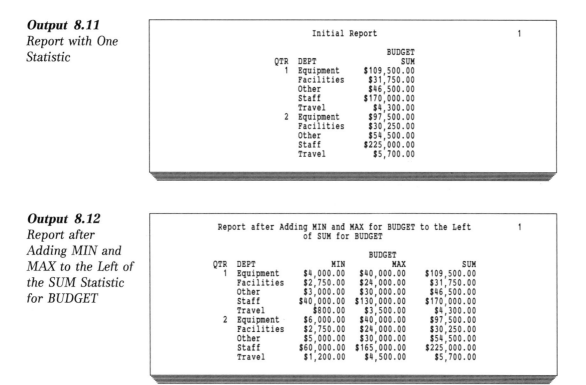

```
                Report after Adding QTR to the Left           1
                             of DEPT

                         QTR  DEPT
                          1   Staff
                          2   Staff
                          1   Staff
                          2   Staff
                          1   Equipment
```

Output 8.11 and Output 8.12 illustrate adding the statistics MIN and MAX for BUDGET to the left of the SUM statistic for BUDGET.

Note: The SUM statistic was added below BUDGET earlier from the STATISTICS window. QTR and DEPT are both group variables.

Output 8.11
Report with One
Statistic

```
                          Initial Report                       1

                                          BUDGET
                     QTR  DEPT                SUM
                      1   Equipment     $109,500.00
                          Facilities     $31,750.00
                          Other          $46,500.00
                          Staff         $170,000.00
                          Travel          $4,300.00
                      2   Equipment      $97,500.00
                          Facilities     $30,250.00
                          Other          $54,500.00
                          Staff         $225,000.00
                          Travel          $5,700.00
```

Output 8.12
Report after
Adding MIN and
MAX to the Left of
the SUM Statistic
for BUDGET

```
            Report after Adding MIN and MAX for BUDGET to the Left    1
                            of SUM for BUDGET

                                          BUDGET
            QTR  DEPT            MIN            MAX            SUM
             1   Equipment    $4,000.00    $40,000.00    $109,500.00
                 Facilities   $2,750.00    $24,000.00     $31,750.00
                 Other        $3,000.00    $30,000.00     $46,500.00
                 Staff       $40,000.00   $130,000.00    $170,000.00
                 Travel         $800.00     $3,500.00      $4,300.00
             2   Equipment    $6,000.00    $40,000.00     $97,500.00
                 Facilities   $2,750.00    $24,000.00     $30,250.00
                 Other        $5,000.00    $30,000.00     $54,500.00
                 Staff       $60,000.00   $165,000.00    $225,000.00
                 Travel       $1,200.00     $4,500.00      $5,700.00
```

Note: You can generate the same report by adding the MIN, MAX, and SUM statistics below BUDGET at the same time.

See Also

the following commands:

□ ADD_ABOVE

□ ADD_BELOW

□ ADD_RIGHT

ADD_LEFT *continued*

the following windows:

□ ADDING

□ COMPUTE

□ COMPUTED VAR

□ DATASET VARS

□ DEFINITION

□ STATISTICS

ADD_RIGHT

Adds one or more data set variables, one or more statistics, or a computed variable to the right of the selected item

---

Action Bar

Edit

Description

The ADD_RIGHT command adds one or more data set variables, one or more statistics, or a computed variable to the right of the selected item.

Usage

1. Select an item.
2. Issue the ADD_RIGHT command.
3. Select **Dataset variable**, **Statistic**, or **Computed variable** from the ADDING window. Depending on your choice, PROC REPORT leads you down one of three paths:

   □ If you select **Dataset variable**, PROC REPORT displays the DATASET VARS window, which lists all the variables in the data set. Select one or more variables from this window. When you exit this window, PROC REPORT opens in turn the DEFINITION window for each variable you selected. In each DEFINITION window, select the characteristics you want for that variable.

   □ If you select **Statistic**, PROC REPORT displays the STATISTICS window. Select one or more statistics from this window.

   In general, when you add a statistic to a report, the column containing the statistic must be associated with a variable; that is, the statistic must either be named in the Statistic field in the DEFINITION window for an analysis variable or appear above or below a display, analysis, or across variable. If you try to add a statistic to the right of a variable, PROC REPORT has no variable to associate with the statistic. The only meaningful selection from the STATISTICS window in such a case is **N**, a simple count of observations, which is independent of any variable. If you select **N**, each value in the new column represents the number of observations described in that row.

If you are adding statistics to the right of another statistic, you can select any statistics. The REPORT procedure associates the new statistics with the variable associated with the selected statistic.

□   If you select **Computed variable**, PROC REPORT displays the COMPUTED VAR window. Enter a name for the computed variable at the prompt. If you are adding a character variable, select the **Character data** check box and, if you wish, specify the length of the variable at the prompt. (The default length is 8.) Then, select the **Edit Program** push button, which opens the COMPUTE window. Use DATA step statements in the COMPUTE window to define the computed variable.

Note: Computed variables exist only in the report. The REPORT procedure does not add them to the data set.

Illustration

Output 8.13 and Output 8.14 illustrate adding the variable DEPT to the right of the variable QTR. DEPT and QTR are both display variables.

Output 8.13
Report with One
Variable

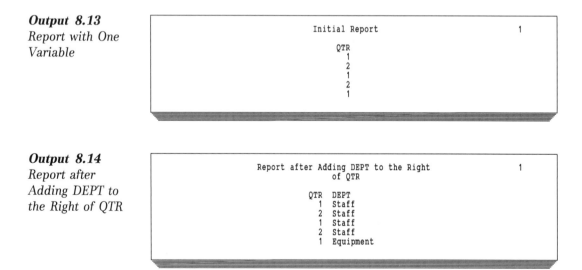

```
                          Initial Report                      1

                              QTR
                               1
                               2
                               1
                               2
                               1
```

Output 8.14
Report after
Adding DEPT to
the Right of QTR

```
                   Report after Adding DEPT to the Right      1
                                of QTR

                      QTR   DEPT
                        1   Staff
                        2   Staff
                        1   Staff
                        2   Staff
                        1   Equipment
```

Output 8.15 and Output 8.16 illustrate adding the statistics MIN and MAX for BUDGET to the right of the SUM statistic for BUDGET.

Note: The SUM statistic was added below BUDGET earlier from the STATISTICS window. QTR and DEPT are both group variables.

ADD_RIGHT *continued*

Output 8.15
Report with One
Statistic

```
                              Initial Report                           1

                                          BUDGET
                 QTR  DEPT                    SUM
                   1  Equipment       $109,500.00
                      Facilities       $31,750.00
                      Other            $46,500.00
                      Staff           $170,000.00
                      Travel            $4,300.00
                   2  Equipment        $97,500.00
                      Facilities       $30,250.00
                      Other            $54,500.00
                      Staff           $225,000.00
                      Travel            $5,700.00
```

Output 8.16
Report after
Adding MIN and
MAX for BUDGET
to the Right of the
SUM Statistic for
BUDGET

```
            Report after Adding MIN and MAX for BUDGET to the Right       1
                              of SUM for BUDGET

                                          BUDGET
             QTR  DEPT              SUM           MIN           MAX
               1  Equipment  $109,500.00    $4,000.00    $40,000.00
                  Facilities  $31,750.00    $2,750.00    $24,000.00
                  Other       $46,500.00    $3,000.00    $30,000.00
                  Staff      $170,000.00   $40,000.00   $130,000.00
                  Travel       $4,300.00      $800.00     $3,500.00
               2  Equipment   $97,500.00    $6,000.00    $40,000.00
                  Facilities  $30,250.00    $2,750.00    $24,000.00
                  Other       $54,500.00    $5,000.00    $30,000.00
                  Staff      $225,000.00   $60,000.00   $165,000.00
                  Travel       $5,700.00    $1,200.00     $4,500.00
```

Note: You can generate the same report by adding the SUM, MIN, and MAX statistics below BUDGET at the same time.

See Also

the following commands:

□ ADD_ABOVE

□ ADD_BELOW

□ ADD_LEFT

the following windows:

□ ADDING

□ COMPUTE

□ COMPUTED VAR

□ DATASET VARS

□ DEFINITION

□ STATISTICS

BREAK

Controls the REPORT procedure's actions at a break (change) in the value of a group or order variable

Action Bar

Edit

Description

The BREAK command controls the REPORT procedure's actions at a break in the value of a group or order variable. Use the BREAK command to perform calculations at a break or to control the location and content of break lines for sets of observations sharing values of a group or order variable. A break line is a line of a report that contains one of the following:

□ characters that visually separate parts of the report.

□ summaries of statistics and computed variables (called a *summary line*).

□ text, values calculated for a set of rows of the report, or both. Chapter 5, "Using Some Advanced Features of the REPORT Procedure," and Chapter 6, "Using the COMPUTE Window," teach you how to use this kind of break line.

When you use the BREAK command, PROC REPORT implements the actions you specify each time the value of the selected group or order variable changes. Thus, break lines produced by the BREAK command summarize or separate sets of observations that share the same value for the selected group or order variable. You can create break lines that summarize the entire report with the RBREAK command.

Usage

1. Select a group or order variable. This variable is the *break variable*.

2. Issue the BREAK command.

3. When the LOCATION window appears on your display, select **After detail** or **Before detail.** If you select **After detail,** the break lines appear after the last row of each set of rows that share the same value of the break variable. If you select **Before detail,** the break lines appear before the first row of each set of rows that share the same value of the break variable.
   Note: *Detail rows* are rows of a report that contain information based on either a single observation in the data set or on a group of observations that have a unique combination of values for all group variables.

4. After you make your selection in the LOCATION window, the REPORT procedure displays the BREAK window. Use the BREAK window to select separator characters, to indicate whether you want to summarize the detail rows, to indicate whether you want to suppress the value of the break variable in the summary line, and to specify a color for all break lines.
   Note: Not all operating systems and devices support all colors, and on some operating systems and devices, one color may map to another color. For

BREAK *continued*

example, if the BREAK window displays BROWN in yellow characters, selecting BROWN results in yellow break lines.

Currently, color appears only when PROC REPORT displays the report in the REPORT window.

5. If you want to add customized break lines or perform calculations, select the **Edit Program** push button from the BREAK window. When the COMPUTE window appears, enter the appropriate statements. You can use most DATA step features in your statements in the COMPUTE window. Use the LINE statement to write text to the report. (See "LINE Statement" in Chapter 10, "The REPORT Language," for details.)

Note: When you attach DATA step statements and LINE statements to a break, the REPORT procedure executes these statements once at each break. Because PROC REPORT calculates statistics for groups before it actually constructs the rows of a report, statistics for sets of rows are available before or after the rows are displayed, as are values for any variables based on these statistics. For instance, if you want to display the sum of the variable BUDGET, add the SUM statistic above or below BUDGET, or specify SUM in the Statistic field in the DEFINITION window for BUDGET. The SUM statistic is then available in the variable BUDGET.SUM at the break, whether the break is before or after the detail rows. You can store this value in another variable whose value won't change as PROC REPORT builds the report (the value of BUDGET.SUM changes in each row) and use it to calculate other values in the report. This technique is particularly useful for calculating percentages, such as the percentage of the budget allocated to each account in a department. For an example of calculating percentages, see "Adding a Customized Break Line" in Chapter 5.

While summary statistics are available independent of the location of the break, some values are not available until the REPORT procedure has constructed the last detail row of a group. For example, you may want to determine the maximum difference between BUDGET and ACTUAL within each department. In such a case, you must compute this difference for each detail row and determine in which row the difference is greatest. The REPORT procedure cannot know which difference is greatest until it has processed the entire set of rows. Therefore, if you want to use the maximum difference in any break lines, the break lines must appear after the detail rows.

Note: Chapter 6 describes in more detail how the REPORT procedure constructs a report.

Illustration

Output 8.17 shows a report without breaks in which DEPT and ACCOUNT are group variables and BUDGET is an analysis variable used to calculate the SUM statistic.

Output 8.17
Report without Breaks

```
                      Report without Breaks                      1

                  DEPT   ACCOUNT       BUDGET
             Equipment     lease   $80,000.00
                           maint   $22,000.00
                        purchase   $60,000.00
                          rental   $10,000.00
                            sets   $15,000.00
                            tape   $20,000.00
            Facilities      rent   $48,000.00
                        supplies    $5,500.00
                           utils    $8,500.00
                 Other    advert   $60,000.00
                        musicfee    $8,000.00
                          talent   $33,000.00
                 Staff  fulltime  $295,000.00
                        parttime  $100,000.00
                Travel       gas    $2,000.00
                          leases    $8,000.00
```

Output 8.18 shows the report produced when you select DEPT as the break variable and place the break lines after each set of rows.

Output 8.18
Placing Break Lines after Sets of Rows

```
                       Breaking after DEPT                       1

                  DEPT   ACCOUNT       BUDGET
             Equipment     lease   $80,000.00
                           maint   $22,000.00
                        purchase   $60,000.00
                          rental   $10,000.00
                            sets   $15,000.00
                            tape   $20,000.00
                        ----------   ----------
             Equipment             $207,000.00
             ----------            ----------
            Facilities      rent   $48,000.00
                        supplies    $5,500.00
                           utils    $8,500.00
                        ----------   ----------
            Facilities             $62,000.00
            ----------            ----------
                 Other    advert   $60,000.00
                        musicfee    $8,000.00
                          talent   $33,000.00
                        ----------   ----------
                 Other             $101,000.00
                 ----------            ----------
                 Staff  fulltime  $295,000.00
                        parttime  $100,000.00
                        ----------   ----------
                 Staff             $395,000.00
                 ----------            ----------
                Travel       gas    $2,000.00
                          leases    $8,000.00
                        ----------   ----------
                Travel             $10,000.00
                ----------            ----------
```

Output 8.19 shows the report produced when you select DEPT as the break variable and place the break lines before each set of rows.

BREAK *continued*

Output 8.19
Placing Break
Lines before Sets
of Rows

```
                        Breaking before DEPT                    1

                   DEPT   ACCOUNT       BUDGET
                 ----------  --------   -----------
                 Equipment             $207,000.00
                 ----------  --------   -----------
                 Equipment     lease    $80,000.00
                               maint    $22,000.00
                            purchase    $60,000.00
                              rental    $10,000.00
                                sets    $15,000.00
                                tape    $20,000.00
                 ----------             -----------
                 Facilities            $62,000.00
                 ----------             -----------
                 Facilities      rent   $48,000.00
                             supplies    $5,500.00
                                utils    $8,500.00
                 ----------             -----------
                     Other            $101,000.00
                 ----------             -----------
                     Other    advert    $60,000.00
                            musicfee     $8,000.00
                              talent    $33,000.00
                 ----------             -----------
                     Staff            $395,000.00
                 ----------             -----------
                     Staff  fulltime  $295,000.00
                            parttime  $100,000.00
                 ----------             -----------
                    Travel             $10,000.00
                 ----------             -----------
                    Travel      gas     $2,000.00
                              leases     $8,000.00
```

Although the locations of the breaks in Output 8.18 and Output 8.19 are different, the selections from the BREAK window that produced the reports are identical:

□ OVERLINE

□ UNDERLINE

□ SUMMARIZE.

See Also

the RBREAK command

the following windows:

□ BREAK

□ COMPUTE

□ LOCATION

CGROW

Increases by one horizontal position the width of the column or columns containing the selected item

Action Bar

Edit

Description

Each time you issue the CGROW command, the REPORT procedure adds one blank character to the width of the columns containing the selected item.

The width of a column can range from 1 to the value of the Linesize attribute in the ROPTIONS window.

Usage

1. Select an item.

2. Issue the CGROW command.

See Also

□ the Width attribute in the DEFINITION window

CSHRINK

Decreases by one horizontal position the width of the column or columns containing the selected item

Action Bar

Edit

Description

Each time you issue the CSHRINK command, the REPORT procedure deletes one horizontal position from the width of the columns containing the selected item.

The width of a column can range from 1 to the value of the Linesize attribute in the ROPTIONS window.

Usage

1. Select an item.

2. Issue the CSHRINK command.

CSHRINK *continued*

See Also

the Width attribute in the DEFINITION window

DEFINE

Displays the characteristics of the selected item

Action Bar

Edit

Description

The DEFINE command opens the DEFINITION window, through which you define the characteristics of an item in the report. These characteristics control

□ the way the REPORT procedure uses the item in the report (for example, as a group variable)

□ the display of values (for example, the format)

□ the display of the entire column containing the item (for example, suppressing the display of a column)

□ the way the REPORT procedure positions values within the column (for example, centering the values)

□ the color of the column header and the values within the column

□ the text of the column header.

Usage

1. Select an item.

2. Issue the DEFINE command.

3. Choose the characteristics from those displayed in the DEFINITION window. For detailed explanations of the choices, see the documentation for the DEFINITION window.

DELETE

Removes an item from the report or a line from a header

Action Bar

Edit

Description

The DELETE command has two functions:

□ removing an item from the report

□ removing one line of a header from the report.

When you delete an item from the report, the REPORT procedure loses all information about that item. Therefore, if you add a data set variable back to the report after deleting it, the REPORT procedure uses the format defined for that variable in the data set, not the format you had assigned to it in the DEFINITION window. For other characteristics, such as the column width and the usage, PROC REPORT uses the defaults. When you add a computed variable back to the report, you must redefine the variable in the COMPUTE window.

Note: The DELETE command has no effect on the data set. Data set variables that you delete from the report remain in the data set.

Usage

To delete an item from the report, follow these steps:

1. Select the item you want to delete. If the column header contains only one line of text, you can select either the header or the header and all values for the item. You select the header and all values by selecting any value of the item. If the header contains more than one line of text, you must select the header and all values.

2. Issue the DELETE command.

To delete a line of text from a header, follow these steps:

1. Select the line of the header that you want to delete.

2. Issue the DELETE command.

When you select one line of a header, PROC REPORT highlights all lines of the header. However, the DELETE command removes only the line of text on which the cursor sits.

LIST

Writes the REPORT language defining the current report to the MESSAGE window

Action Bar

Locals

Description

Each time you issue the LIST command, the REPORT procedure writes the REPORT language defining the current report to the MESSAGE window. You can use this command in conjunction with Chapter 10 to learn the REPORT language.

Usage

Issue the LIST command.

See Also

the MESSAGE window

MESSAGE

Opens the MESSAGE window, which contains notes, warnings, and errors issued by the REPORT procedure, as well as the REPORT language generated by the LIST command.

Action Bar

Locals

Description

The MESSAGE command opens the MESSAGE window, which displays notes, warnings, and errors issued by the REPORT procedure, as well as the REPORT language generated by the LIST command.

Usage

Issue the MESSAGE command.

MOVE

Moves an item from one location in the report to another

Action Bar

Edit

Description

The MOVE command moves the selected item from one location in the report to another.

Usage

1. Select the item you want to move.

2. Issue the MOVE command.

3. Select the item around which you want to move the first item you selected.

4. Issue the appropriate command: ADD_ABOVE, ADD_BELOW, ADD_LEFT, or ADD_RIGHT.

Illustration

Output 8.20 and Output 8.21 illustrate moving the variable BUDGET to a position below the variable QTR. In Output 8.20, DEPT and ACCOUNT are group variables; QTR is an across variable; and BUDGET is an analysis variable used to calculate the SUM statistic. Because the report contains no variable below QTR, the REPORT procedure displays the value of the N statistic, the number of observations for each quarter that share a value of DEPT and ACCOUNT. The values for BUDGET represent the sum of BUDGET for both quarters.

Output 8.20
Before Moving
BUDGET below
QTR

```
                        Before Moving BUDGET below QTR                    1
                                        QTR
                DEPT    ACCOUNT     1          2            BUDGET
                Equipment  lease    1          1         $80,000.00
                           maint    1          1         $22,000.00
                        purchase    1          1         $60,000.00
                          rental    1          1         $10,000.00
                            sets    1          1         $15,000.00
                            tape    1          1         $20,000.00
                Facilities  rent    1          1         $48,000.00
                        supplies    1          1          $5,500.00
                           utils    1          1          $8,500.00
                    Other  advert    1          1         $60,000.00
                        musicfee    1          1          $8,000.00
                          talent    1          1         $33,000.00
                    Staff fulltime   1          1        $295,000.00
                        parttime    1          1        $100,000.00
                    Travel   gas    1          1          $2,000.00
                          leases    1          1          $8,000.00
```

MOVE *continued*

Output 8.21 shows the results of moving BUDGET below QTR. The values for BUDGET now represent the values for each quarter.

Output 8.21
After Moving
BUDGET below
QTR

```
          After Moving BUDGET below QTR                    1

                              QTR
                          1                   2
        DEPT  ACCOUNT       BUDGET          BUDGET
   Equipment    lease   $40,000.00      $40,000.00
                maint   $10,000.00      $12,000.00
             purchase   $40,000.00      $20,000.00
               rental    $4,000.00       $6,000.00
                 sets    $7,500.00       $7,500.00
                 tape    $8,000.00      $12,000.00
  Facilities     rent   $24,000.00      $24,000.00
             supplies    $2,750.00       $2,750.00
                utils    $5,000.00       $3,500.00
       Other   advert   $30,000.00      $30,000.00
             musicfee    $3,000.00       $5,000.00
               talent   $13,500.00      $19,500.00
       Staff fulltime  $130,000.00     $165,000.00
             parttime   $40,000.00      $60,000.00
      Travel      gas      $800.00       $1,200.00
               leases    $3,500.00       $4,500.00
```

When you move a statistic, the REPORT procedure calculates the statistic for the variable associated with the new position. For example, consider the reports in Output 8.22 and Output 8.23. In these reports DEPT is a group variable, and BUDGET and ACTUAL are display variables with statistics added below them. These reports illustrate moving the statistic MAX (for BUDGET) from the left-hand to the right-hand side of the statistic MIN (also for BUDGET).

Output 8.22
Report before
Moving the MAX
Statistic

```
         Report with MAX for BUDGET to the Left            1
                     of MIN for BUDGET

                       BUDGET                ACTUAL
   DEPT              MAX          MIN            MIN
   Equipment  $40,000.00    $4,000.00     $3,998.87
   Facilities $24,000.00    $2,750.00     $2,216.55
   Other      $30,000.00    $3,000.00     $2,550.50
   Staff     $165,000.00   $40,000.00    $43,850.12
   Travel     $4,500.00      $800.00       $537.26
```

Output 8.23
Report after
Moving the MAX
Statistic

```
      Report after Moving MAX for BUDGET to the Right       1
                     of MIN for BUDGET

                       BUDGET                ACTUAL
   DEPT              MIN          MAX            MIN
   Equipment   $4,000.00   $40,000.00     $3,998.87
   Facilities  $2,750.00   $24,000.00     $2,216.55
   Other       $3,000.00   $30,000.00     $2,550.50
   Staff      $40,000.00  $165,000.00    $43,850.12
   Travel       $800.00    $4,500.00       $537.26
```

Because the MAX statistic applies to BUDGET in both reports, the numbers in the column for MAX don't change when you move the statistic.

Now consider starting with the report in Output 8.22 and moving MAX from its original position under BUDGET to the right of MIN under ACTUAL. In this case, the MAX statistic applies to BUDGET before the move and to ACTUAL after the move. Therefore, as you can see by comparing Output 8.24 to Output 8.23, the numbers in the column for MAX do change.

Output 8.24
Moving a Statistic below a Different Variable

```
                 Report after Moving MAX for BUDGET to the Right          1
                              of MIN for ACTUAL

                           BUDGET            ACTUAL
            DEPT              MIN               MIN           MAX
            Equipment     $4,000.00        $3,998.87     $48,282.38
            Facilities    $2,750.00        $2,216.55     $24,000.00
            Other         $3,000.00        $2,550.50     $37,325.64
            Staff        $40,000.00       $43,850.12    $166,345.75
            Travel          $800.00          $537.26      $3,889.65
```

See Also

the following commands:

□ ADD_ABOVE

□ ADD_BELOW

□ ADD_LEFT

□ ADD_RIGHT

PAGE

Displays the next page of the report

Action Bar

View

Description

The REPORT procedure breaks the rows and columns of a report into pages in accordance with the page size and line size. Figure 8.1 shows the pages of a report that contains 15 columns and 40 rows in an environment where the page size and line size allow 5 columns and 20 rows to fit on one page. Each time you issue the PAGE command, PROC REPORT displays the next page of the report.

PAGE *continued*

Figure 8.1
Pages of a Report

Note: If the page size is greater than the length of the window, which it is on many systems by default, use the SAS Display Manager commands FORWARD and BACKWARD to view the entire page. If the line size is greater than the width of the window, which it is on many systems by default, use the SAS Display Manager commands RIGHT and LEFT to view the entire page. These commands are in the pull-down menu under **View** in the action bar at the top of the REPORT window.

Usage

Issue the PAGE command.

PAGEBACK

Displays the previous page of the report

Action Bar

View

Description

The REPORT procedure breaks the rows and columns of a report into pages in accordance with the page size and line size. Figure 8.2 shows the pages of a report that contains 15 columns and 40 rows in an environment where the page size and line size allow 5 columns and 20 rows to fit on one page. Each time you issue the PAGEBACK command, PROC REPORT displays the previous page of the report.

Figure 8.2
Pages of a Report

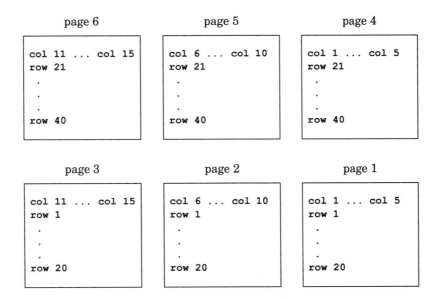

Note: If the page size is greater than the length of the window, which it is on many systems by default, use the display manager commands FORWARD and BACKWARD to view the entire page. If the line size is greater than the width of the window, which it is on many systems by default, use the display manager commands RIGHT and LEFT to view the entire page. These commands are in the pull-down menu under **View** in the action bar at the top of the REPORT window.

Usage

Issue the PAGEBACK command.

PROFILE

Enables you to customize certain aspects of the REPORT procedure

Action Bar

Locals

Description

Opens the PROFILE window, which enables you to customize the action bars and pull-down menus in the REPORT and COMPUTE windows, to set defaults for selected options, and to change the colors of most REPORT windows. For detailed explanations of these features, see the documentation for the PROFILE window.

Usage

Issue the PROFILE command.

QUIT

Terminates the REPORT procedure

Action Bar

File

Description

The QUIT command terminates the report procedure.

Usage

Issue the QUIT command.

RBREAK

Controls the REPORT procedure's actions at a break at the beginning or end of the report

Action Bar

Edit

Description

The RBREAK command controls the REPORT procedure's actions at a break at the beginning or end of the report. Use the RBREAK command to perform calculations at a break or to control the location and content of break lines that apply to the entire report. A break line is a line of a report that contains one of the following:

□ characters that visually separate parts of the report.

□ summaries of statistics and computed variables (called a *summary line*).

□ text, values calculated for a set of rows of the report (in this case, all rows of the report), or both. Chapter 5 and Chapter 6 teach you how to use this kind of break line.

When you use the RBREAK command, PROC REPORT implements the actions you specify at the beginning or end of the report. (Use the BREAK command to create break lines that summarize or separate sets of observations that share a value for the selected group or order variable.)

Usage

1. Issue the RBREAK command.

2. When the LOCATION window appears on your display, select **After detail** or **Before detail**. If you select **After detail**, the break lines appear at the end of the report. If you select **Before detail**, the break lines appear at the beginning of the report.

   Note: *Detail rows* are lines of a report that contain information based on either a single observation in the data set or on a group of observations that have a unique combination of values for all group variables.

3. After you make your selection in the LOCATION window, the REPORT procedure displays the BREAK window. Use the BREAK window to select separator characters, to indicate whether you want to summarize the detail rows, and to specify a color for all break lines.

   Note: Not all operating systems and devices support all colors, and on some operating systems and devices, one color may map to another color. For example, if the BREAK window displays BROWN in yellow characters, selecting BROWN results in yellow break lines.

   Currently, color appears only when PROC REPORT displays the report in the REPORT window.

4. If you want to add customized break lines, select the **Edit Program** push button from the BREAK window. When the COMPUTE window appears, enter the appropriate statements. You can use most DATA step features in your statements in the COMPUTE window. Use the LINE statement to write text to the report. (See Chapter 10 for details.)

   Note: When you attach DATA step statements and LINE statements to a break, the REPORT procedure executes these statements once at each break in the report. Because PROC REPORT calculates statistics for the report before it actually constructs the rows of a report, statistics are available at the beginning or end of the report, as are values for any variables based on these statistics. For instance, if you want to display the sum of the variable BUDGET, add the SUM statistic above or below BUDGET, or specify `SUM` in the Statistic field in the DEFINITION window for BUDGET. The SUM statistic is then available in the variable BUDGET.SUM at the break, whether the break is at the beginning or end of the report. You can store this value in another variable whose value won't change as PROC REPORT builds the report (the value of BUDGET.SUM changes in each row) and use it to calculate other values in the report. This technique is particularly useful for calculating percentages, such as the percentage of the budget allocated to each account in a department. For an example of calculating percentages, see "Adding a Customized Break Line" in Chapter 5.

   While summary statistics are available independent of the location of the break, some values are not available until the REPORT procedure has constructed the last detail row. For example, you may want to determine the maximum difference between BUDGET and ACTUAL for the entire company. In such a case, you must compute this difference for each detail row and determine in which row the difference is greatest. The REPORT procedure cannot know which difference is greatest until it has processed all rows. Therefore, if you want to use the maximum difference in any break lines, the break lines must appear at the end of the report. Chapter 6 describes in more detail how the REPORT procedure constructs a report.

RBREAK *continued*

Illustration

Output 8.25 shows a report with break lines at the beginning of the report. Output 8.26 shows the same report with the break lines at the end of the report.

Although the locations of the breaks in Output 8.25 and Output 8.26 are different, the selections from the BREAK window that produced the reports are identical:

□   DOUBLE OVERLINE

□   DOUBLE UNDERLINE

□   SUMMARIZE.

Output 8.25
Break Lines at the
Beginning of a
Report

```
          Creating a Break at the Beginning of a Report              1

            DEPT         ACCOUNT        BUDGET
            ==========   ========    ===========
                                     $775,000.00

            ==========   ========    ===========
            Equipment    lease        $80,000.00
                         maint        $22,000.00
                         purchase     $60,000.00
                         rental       $10,000.00
                         sets         $15,000.00
                         tape         $20,000.00
            Facilities   rent         $48,000.00
                         supplies      $5,500.00
                         utils         $8,500.00
            Other        advert       $60,000.00
                         musicfee      $8,000.00
                         talent       $33,000.00
            Staff        fulltime    $295,000.00
                         parttime    $100,000.00
            Travel       gas           $2,000.00
                         leases        $8,000.00
```

Output 8.26
Break Lines at the
End of a Report

```
          Creating a Break at the End of a Report                    1

            DEPT         ACCOUNT        BUDGET
            Equipment    lease        $80,000.00
                         maint        $22,000.00
                         purchase     $60,000.00
                         rental       $10,000.00
                         sets         $15,000.00
                         tape         $20,000.00
            Facilities   rent         $48,000.00
                         supplies      $5,500.00
                         utils         $8,500.00
            Other        advert       $60,000.00
                         musicfee      $8,000.00
                         talent       $33,000.00
            Staff        fulltime    $295,000.00
                         parttime    $100,000.00
            Travel       gas           $2,000.00
                         leases        $8,000.00
                                     ===========
                                     $775,000.00
                                     ===========
```

See Also

the RBREAK command

the following windows:

- □ BREAK
- □ COMPUTE
- □ LOCATION

REFRESH

Displays the most recent version of the report

Action Bar

Edit

Description

The REFRESH command displays the most recent version of the report. By default, PROC REPORT updates the report each time you redefine the report by adding or deleting an item, by changing information in the DEFINITION window, or by changing information in the BREAK window. (You can override this behavior by selecting the DEFER option from the ROPTIONS window.) However, you must issue the REFRESH command (or make a change to the report that causes PROC REPORT to issue the command) in order to see changes in the following circumstances:

- □ when you change a header
- □ when you change the title of the report or footnotes
- □ when you do anything with the DEFER option in effect.

Usage

Issue the REFRESH command.

See Also

the DEFER option in the ROPTIONS window

RKEYS

Displays the RKEYS window if it is not currently displayed and refreshes the definitions in the RKEYS window with the most recently created ones

Action Bar

Locals

Description

The RKEYS command displays the RKEYS window if it is not currently displayed and refreshes the definitions in the window with the most recently created ones.

For information on changing the default function key definitions, see "Defining a Function Key" in Chapter 7, "Using the REPORT Procedure in a Windowing Environment."

Usage

Issue the RKEYS command.

See Also

the NORKEYS option in the PROC REPORT statement

RLOAD

Loads the specified report definition or, if you do not specify an argument, displays a selection of report definitions

Action Bar

File

Description

The RLOAD command loads a stored report definition, which it applies to the current data set. The syntax for this command enables you to choose between opening the RLOAD window, which displays a selection of report definitions, and directly specifying the definition to load:

RLOAD <*libref.catalog.entry*>

where *libref* points to the SAS data library that contains the catalog in which the report definition resides; *catalog* identifies the catalog that contains the report definition; and *entry* is the name of the entry that is the report definition.

Usage

1. Issue the RLOAD command. If you want to specify a report definition directly rather than opening the RLOAD window, you must issue the command from the command line or from a function key whose definition includes the appropriate arguments (see "Issuing Commands from the REPORT Procedure," in Chapter 7).

2. If you do not specify a report definition, the RLOAD command opens the RLOAD window. Enter a libref and a catalog name in the fields in the RLOAD window and press ENTER. PROC REPORT lists all report definitions (entries of type REPT) in the specified catalog.

3. Select the report definition you want to use.

4. Exit the RLOAD window.

See Also

the RLOAD window

the RSTORE command

the OUTREPT= option and the REPORT= option in Chapter 7, "Using the REPORT Procedure in a Windowing Environment"

ROPTIONS

Opens the ROPTIONS window, which controls the layout and display of the entire report and identifies the SAS data library and catalog containing CBT or HELP entries for items in the report

Action Bar

Locals

Description

The ROPTIONS command opens the ROPTIONS window, through which you control the layout and display of the entire report and identify the SAS data library and catalog containing HELP entries for items in the report.

Usage

1. Issue the ROPTIONS command.

2. Select the settings from those displayed in the ROPTIONS window. For detailed explanations of the choices, see the documentation for the ROPTIONS window.

RSTORE

Stores a report definition in the specified catalog entry or, if you do not specify an argument, opens the RSTORE window, which prompts you for a catalog entry

Action Bar

File

Description

The RSTORE command stores a report definition. The REPORT procedure stores a report definition in a catalog entry of type REPT. You can apply this report definition to any SAS data set that contains the variables you used in the report definition. To use the definition, issue the RLOAD command or use the REPORT= option when you invoke the REPORT procedure.

The syntax for this command enables you to choose between opening the RSTORE window, which displays a selection of report definitions, and directly specifying the catalog entry in which to store the report definition:

RSTORE <*libref.catalog.entry*>

where *libref* points to the SAS data library that contains the catalog in which to store the report definition; *catalog* identifies the catalog to contain the report definition; and *entry* is the name of the entry that is the report definition.

Usage

□ Issue the RSTORE command. If you want to specify a catalog entry in which to store the report definition directly rather than opening the RSTORE window, you must issue the command from the command line or from a function key whose definition includes the appropriate arguments (see "Issuing Commands from the REPORT Procedure," in Chapter 7).

□ If you do not specify a catalog entry, the RSTORE command opens the RSTORE window. You must enter a libref, catalog name, and entry name in the fields in the RSTORE window. Although it is not necessary, you can also enter a description of the definition in the Description field. This description is useful because it appears with the name of the definition in the RLOAD window and the CATALOG window, helping you to identify the definition.

▶ *Caution* . . . . *If the catalog entry you specify already exists, the REPORT procedure writes over it without warning.* ▲

See Also

the OUTREPT= option in Chapter 7, "Using the REPORT Procedure in a Windowing Environment"

the RLOAD command

the RSTORE window

SPAN

Produces a header that spans multiple columns

Action Bar

Edit

Description

The SPAN command produces a header that spans multiple columns.

Usage

1. Select the leftmost item that you want the header to cover.

2. Issue the SPAN command.

3. Select the rightmost item that you want the header to cover.

4. Issue the ADD_ABOVE command. PROC REPORT highlights a blank line that spans the columns you selected.

5. Enter text for the header on the highlighted line. If you need more lines for the header, use the ADD_ABOVE or ADD_BELOW command to add empty header lines.

   If the first and last characters of a header are —, =, _, ., *, or +, or if the first and last characters of a header are < and > or > and <, the REPORT procedure expands the header by repeating the first and last characters to fill the space over the columns the heading spans.

Illustration

The report in Output 8.27 contains a header that spans the three columns containing figures.

Output 8.27
Using a Header
That Spans
Multiple Columns

```
Using a Header that Spans BUDGET, ACTUAL, and BALANCE          1

                         Figures Are Rounded to the
                              Nearest Dollar
                       --------------------------------

                              Amount    Amount
            Department  Account  Budgeted    Spent   Balance
            ----------  --------  ---------  ---------  ---------

            Equipment   lease    $80,000   $80,000        $0
                        maint    $22,000   $18,217    $3,783
                        purchase $60,000   $66,052   $-6,052
                        rental   $10,000    $9,482      $518
                        sets     $15,000   $16,422   $-1,422
                        tape     $20,000   $18,256    $1,744
            ----------           ---------  ---------  ---------
            Equipment           $207,000  $208,429   $-1,429
            ----------           ---------  ---------  ---------
```

SPAN *continued*

Facilities	rent	$48,000	$48,000	$0
	supplies	$5,500	$4,959	$541
	utils	$8,500	$7,668	$832
Facilities		$62,000	$60,627	$1,373
Other	advert	$60,000	$69,803	$-9,803
	musicfee	$8,000	$7,426	$574
	talent	$33,000	$31,411	$1,589
Other		$101,000	$108,640	$-7,640
Staff	fulltime	$295,000	$293,988	$1,012
	parttime	$100,000	$99,869	$131
Staff		$395,000	$393,858	$1,142
Travel	gas	$2,000	$1,522	$478
	leases	$8,000	$6,935	$1,065
Travel		$10,000	$8,457	$1,543
		$775,000	$780,011	$-5,011

See Also

the following commands:

- □   ADD_ABOVE
- □   ADD_BELOW

Chapter 9 REPORT Windows

ADDING

Displays the kinds of items you can add to a report

Display

```
┌REPORT─────────────────────────────────────────────────────────┐
│File Edit Search View Locals Globals                            │
│                                                                │
│                       The SAS System                         1 │
│                                                                │
│              DEPT        ACCOUNT       BUDGET                   │
│              Equipment   lease       $80,000.00                │
│                          maint       $22,000.00                │
│                          purchase    $60,0┌ADDING──────────────┐│
│                          rental      $10,0│...above BUDGET     ││
│                          sets        $15,0│                    ││
│                          tape        $20,0│ o Dataset variable ││
│              Facilities  rent        $48,0│ o Statistic        ││
│                          supplies     $5,5│ o Header line      ││
│                          utils        $8,5│                    ││
│              Other       advert      $60,0│ <OK>     <Cancel>  ││
│                          musicfee     $8,0│                    ││
│                          talent      $33,0└────────────────────┘│
│                                                                │
├RKEYS───────────────────────────────────────────────────────────┤
│Add_above <F1>     Add_below <F2>     Add_left  <F3>     Add_right <F4>│
│Break     <F5>     CGrow     <F6>     CShrink   <F7>     Define    <F8>│
│Delete    <F9>     Move      <SHF F0> Page      <SHF F1> Pageback  <SHF F2>│
└──────────────────────────────────────────────────────────R─────┘
```

ADDING *continued*

```
┌REPORT────────────────────────────────────────────────────────────────┐
│File Edit Search View Locals Globals                                   │
│                                                                       │
│                         The SAS System                              1 │
│                                                                       │
│               DEPT        ACCOUNT       BUDGET                        │
│               Equipment   lease       $80,000.00                      │
│                           maint       $22,000.00                      │
│                           purchase    $60,0┌ADDING────────────────────┐│
│                           rental      $10,0│...to the right of BUDGET ││
│                           sets        $15,0│                          ││
│                           tape        $20,0│   o Dataset variable     ││
│               Facilities  rent        $48,0│   o Statistic            ││
│                           supplies     $5,5│   o Computed variable    ││
│                           utils        $8,5│                          ││
│               Other       advert      $60,0│   <OK>      <Cancel>     ││
│                           musicfee     $8,0│                          ││
│                           talent      $33,0└──────────────────────────┘│
│                                                                       │
├RKEYS──────────────────────────────────────────────────────────────────┤
│Add_above <F1>      Add_below <F2>      Add_left  <F3>     Add_right <F4>│
│Break     <F5>      CGrow     <F6>      CShrink   <F7>     Define    <F8>│
│Delete    <F9>      Move    <SHF F0>    Page    <SHF F1>   Pageback <SHF F2>│
└──────────────────────────────────────────────────────────────────R────┘
```

Access

Issue one of the following commands:

□ ADD_ABOVE

□ ADD_BELOW

□ ADD_RIGHT

□ ADD_LEFT.

Description

Use the ADDING window to add data set variables, computed variables, or statistics to the report, or to add a blank line to a header. The ADDING window contains text that tells you the name of the selected item and how the item you are adding relates to it spatially (above, below, right, or left).

If you open the window with the ADD_ABOVE or ADD_BELOW command, the choices in the window are **Dataset variable, Statistic,** and **Header line.** If you open the window with the ADD_RIGHT or ADD_LEFT command, the choice **Computed variable** replaces **Header line.** The choices are explained below:

Dataset variable

adds one or more data set variables to the report. When you select **Dataset variable,** the REPORT procedure displays the DATASET VARS window, which lists all the variables in the data set.

Statistic

adds one or more statistics to the report. When you select **Statistic,** PROC REPORT displays the STATISTICS window, which lists all the statistics you can add to the report.

Header line

adds a blank line to the selected header. You can type whatever text you want in this space.

Computed variable

adds a new variable to the report. When you select **Computed variable,** PROC REPORT displays the COMPUTED VAR window, which prompts you for the name of the variable you want to add. If the variable is a character variable, you can also specify its length.

   Note: The REPORT procedure does not add computed variables to the data set. They exist only in the report.

BREAK

Presents choices controlling the REPORT procedure's actions at breaks (changes) in the values of an order or group variable or at the beginning or end of the report

Display

```
┌REPORT─────────────────────────────────────────────────────────────────────
 File Edit Search View Locals Globals

                        The SAS System                                    1

                ┌BREAK───────────────────────────────────────┐
                │         Breaking AFTER   DEPT               │
                │  Options                        Color       │
                │   _ OVERLINE                    BLUE        │
                │   _ DOUBLE OVERLINE   (=)       RED         │
                │   _ UNDERLINE                   PINK        │
                │   _ DOUBLE UNDERLINE (=)        GREEN       │
                │                                 CYAN        │
                │   _ SKIP                        YELLOW      │
                │   _ PAGE                        WHITE       │
                │                                 ORANGE      │
                │   _ SUMMARIZE                   BLACK       │
                │   _ SUPPRESS                    MAGENTA     │
┌RKEYS─────────┐│                                 GRAY        │
 Add_above <F1>│A                                 BROWN       │───────────┐
 Break     <F5>│C  <Edit Program>   <OK>    <Cancel>          │Add_right <F4>
 Delete    <F9>│M                                            │Define    <F8>
               └└────────────────────────────────────────────┘Pageback  <SHF F2>
                                                               └─────────R───
```

Access

Issue one of the following commands:

□ BREAK

□ RBREAK.

Note: In both cases, you first pass through the LOCATION window.

Description

Contents of Break Lines

The choices you make in the BREAK window control the contents of break lines in your report. A *break line* is a line of a report that contains one of the following:

□ characters that visually separate parts of the report.

□ summaries of statistics and computed variables (called a *summary line*).

□ text, values calculated for a set of rows of the report, or both. Chapter 5, "Using Some Advanced Features of the REPORT Procedure," and Chapter 6, "Using the COMPUTE Window," explain how to use this kind of break line.

Location of Break Lines

You can create breaks in a report

□   at the beginning or end of the report

□   each time the value of the selected group or order variable (the *break variable*)
changes.

A header in the BREAK window describes the location of the break lines you
create. For instance, consider a report with a column that contains the names of
departments in a company. To create a break line each time the value of DEPT
changes, select **DEPT** and issue the BREAK command. If you select **After detail**
from the LOCATION window, the header in the BREAK window reads

```
Breaking AFTER DEPT
```

You can open the BREAK window with either the BREAK command or the
RBREAK command. If you open the BREAK window with the BREAK command,
the break lines you define appear in the report each time the value of the break
variable changes. In this case, summary lines summarize the statistics and
computed variables for each set of detail rows that have the same value for the
break variable. By default, the summary lines also include the current value of the
break variable.

Your selection in the LOCATION window determines whether the break lines
appear before the first or after the last row of each set of detail rows.

Note: Summary statistics are available either before or after a break because
PROC REPORT calculates the statistics before it begins constructing the rows of
the report.

Consider the report in Output 9.1. In this report DEPT and ACCOUNT are
group variables. BUDGET is a display variable with the statistics MIN, MAX, and
MEAN added below it. The report uses DEPT as a break variable. Each summary
line in this report contains

□   the current value of DEPT

□   the MIN of BUDGET calculated for all rows having the same value of DEPT

□   the MAX of BUDGET calculated for all rows having the same value of DEPT

□   the MEAN of BUDGET calculated for all rows having the same value of DEPT.

BREAK *continued*

Output 9.1
Creating a Break
on a Variable

```
        Creating a Break Each Time the Value of DEPT Changes            1

                                       BUDGET
         DEPT        ACCOUNT        MIN          MAX          MEAN
         Equipment   lease      $40,000.00   $40,000.00   $40,000.00
                     maint      $10,000.00   $12,000.00   $11,000.00
                     purchase   $20,000.00   $40,000.00   $30,000.00
                     rental      $4,000.00    $6,000.00    $5,000.00
                     sets        $7,500.00    $7,500.00    $7,500.00
                     tape        $8,000.00   $12,000.00   $10,000.00
         ----------             ----------   ----------   ----------
         Equipment              $4,000.00    $40,000.00   $17,250.00
         ----------             ----------   ----------   ----------

         Facilities  rent       $24,000.00   $24,000.00   $24,000.00
                     supplies    $2,750.00    $2,750.00    $2,750.00
                     utils       $3,500.00    $5,000.00    $4,250.00
         ----------             ----------   ----------   ----------
         Facilities             $2,750.00    $24,000.00   $10,333.33
         ----------             ----------   ----------   ----------

         Other       advert     $30,000.00   $30,000.00   $30,000.00
                     musicfee    $3,000.00    $5,000.00    $4,000.00
                     talent     $13,500.00   $19,500.00   $16,500.00
         ----------             ----------   ----------   ----------
         Other                  $3,000.00    $30,000.00   $16,833.33
         ----------             ----------   ----------   ----------

         Staff       fulltime  $130,000.00  $165,000.00  $147,500.00
                     parttime   $40,000.00   $60,000.00   $50,000.00
         ----------             ----------   ----------   ----------
         Staff                  $40,000.00  $165,000.00   $98,750.00
         ----------             ----------   ----------   ----------

         Travel      gas          $800.00    $1,200.00    $1,000.00
                     leases      $3,500.00    $4,500.00    $4,000.00
         ----------             ----------   ----------   ----------
         Travel                   $800.00    $4,500.00    $2,500.00
         ----------             ----------   ----------   ----------
```

If the report included computed variables, the REPORT procedure would calculate their values in the summary lines from the other values in the summary lines.

If you open the BREAK window with the RBREAK command, the break lines you define appear at either the beginning or the end of the report, depending on your choice in the LOCATION window. The report in Output 9.2, for example, includes a break at the end of the report. Again, DEPT and ACCOUNT are group variables, and BUDGET is a display variable with the statistics MIN, MAX, and MEAN added below it. In this case, the summary line summarizes the statistics for the entire report. If the report included computed variables, the REPORT procedure would calculate their values in the summary line from the other values in the summary line.

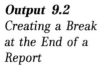

Output 9.2
*Creating a Break
at the End of a
Report*

```
                     Creating a Break at the End of a Report                1
                                         BUDGET
          DEPT       ACCOUNT        MIN          MAX          MEAN
          Equipment  lease      $40,000.00   $40,000.00   $40,000.00
                     maint      $10,000.00   $12,000.00   $11,000.00
                     purchase   $20,000.00   $40,000.00   $30,000.00
                     rental      $4,000.00    $6,000.00    $5,000.00
                     sets        $7,500.00    $7,500.00    $7,500.00
                     tape        $8,000.00   $12,000.00   $10,000.00
          Facilities rent       $24,000.00   $24,000.00   $24,000.00
                     supplies    $2,750.00    $2,750.00    $2,750.00
                     utils       $3,500.00    $5,000.00    $4,250.00
          Other      advert     $30,000.00   $30,000.00   $30,000.00
                     musicfee    $3,000.00    $5,000.00    $4,000.00
                     talent     $13,500.00   $19,500.00   $16,500.00
          Staff      fulltime  $130,000.00  $165,000.00  $147,500.00
                     parttime   $40,000.00   $60,000.00   $50,000.00
          Travel     gas           $800.00    $1,200.00    $1,000.00
                     leases      $3,500.00    $4,500.00    $4,000.00
                                -----------  -----------  -----------
                                    $800.00  $165,000.00   $24,218.75
                                -----------  -----------  -----------
```

As Output 9.3 shows, you can create multiple breaks in a single report. This report contains two kinds of breaks: one uses DEPT as a break variable; the other appears at the end of the report.

Output 9.3
*Creating Multiple
Kinds of Breaks in
a Single Report*

```
                       Creating a Break after DEPT                          1
                                     and
                 Creating a Break at the End of the Report

                                         BUDGET
          DEPT       ACCOUNT        MIN          MAX          MEAN
          Equipment  lease      $40,000.00   $40,000.00   $40,000.00
                     maint      $10,000.00   $12,000.00   $11,000.00
                     purchase   $20,000.00   $40,000.00   $30,000.00
                     rental      $4,000.00    $6,000.00    $5,000.00
                     sets        $7,500.00    $7,500.00    $7,500.00
                     tape        $8,000.00   $12,000.00   $10,000.00
          ----------            -----------  -----------  -----------
          Equipment              $4,000.00   $40,000.00   $17,250.00
          ----------            -----------  -----------  -----------

          Facilities rent       $24,000.00   $24,000.00   $24,000.00
                     supplies    $2,750.00    $2,750.00    $2,750.00
                     utils       $3,500.00    $5,000.00    $4,250.00
          ----------            -----------  -----------  -----------
          Facilities            $2,750.00   $24,000.00   $10,333.33
          ----------            -----------  -----------  -----------

          Other      advert     $30,000.00   $30,000.00   $30,000.00
                     musicfee    $3,000.00    $5,000.00    $4,000.00
                     talent     $13,500.00   $19,500.00   $16,500.00
          ----------            -----------  -----------  -----------
          Other                 $3,000.00   $30,000.00   $16,833.33
          ----------            -----------  -----------  -----------

          Staff      fulltime  $130,000.00  $165,000.00  $147,500.00
                     parttime   $40,000.00   $60,000.00   $50,000.00
          ----------            -----------  -----------  -----------
          Staff                $40,000.00  $165,000.00   $98,750.00
          ----------            -----------  -----------  -----------

          Travel     gas           $800.00    $1,200.00    $1,000.00
                     leases      $3,500.00    $4,500.00    $4,000.00
          ----------            -----------  -----------  -----------
          Travel                  $800.00    $4,500.00    $2,500.00
          ----------            -----------  -----------  -----------

                                ===========  ===========  ===========
                                    $800.00  $165,000.00   $24,218.75
                                ===========  ===========  ===========
```

BREAK *continued*

Options

Your selection of options in the BREAK window determines the contents of your break lines. Detailed descriptions of each choice follow.

OVERLINE

writes a line of hyphens (-) across the report above each value that appears in the summary line or that would appear in the summary line if you selected the **SUMMARIZE** option from the BREAK window.

DOUBLE OVERLINE

writes a line of equals signs (=) across the report above each value that appears in the summary line or that would appear in the summary line if you selected the **SUMMARIZE** option from the BREAK window. If you select both **OVERLINE** and **DOUBLE OVERLINE**, PROC REPORT uses **OVERLINE**.

UNDERLINE

writes a line of hyphens (-) across the report below each value that appears in the summary line or that would appear in the summary line if you selected the **SUMMARIZE** option from the BREAK window.

DOUBLE UNDERLINE

writes a line of equals signs (=) across the report below each value that appears in the summary line or that would appear in the summary line if you selected the **SUMMARIZE** option from the BREAK window. If you select both **UNDERLINE** and **DOUBLE UNDERLINE**, PROC REPORT uses **UNDERLINE**.

SKIP

writes a blank line after the last line of each group of break lines.

PAGE

starts a new page after the last line of each group of break lines.

SUMMARIZE

writes a summary line in each group of break lines.

SUPPRESS

suppresses the printing of the current value of the break variable and any underlining and overlining of that value in the summary line. If you open the BREAK window with the RBREAK command, the SUPPRESS option has no effect because a break at the beginning or end of the report does not use a break variable.

Color

selects the color of the break lines. Available colors follow:

BLUE	**GREEN**	**WHITE**	**MAGENTA**
RED	**CYAN**	**ORANGE**	**GRAY**
PINK	**YELLOW**	**BLACK**	**BROWN**

Note: Not all operating systems and devices support all colors, and on some operating systems and devices, one color may map to another color. For example, if the BREAK window displays **BROWN** in yellow characters, selecting **BROWN** results in yellow break lines.

Currently, color appears only when PROC REPORT displays the report in the REPORT window.

When a break contains more than one break line, the order in which the break lines appear is as follows:

1. overlining or double overlining

2. summary line

3. underlining or double underlining

4. break lines defined in the COMPUTE window for that break

5. skipped line.

Output 9.4 shows a report with a break after DEPT. In this report DEPT (the break variable) and ACCOUNT are both group variables, and BUDGET is an analysis variable used to calculate the SUM statistic. The options chosen from the BREAK window to create this report follow:

1. **DOUBLE OVERLINE**

2. **DOUBLE UNDERLINE**

3. **PAGE**

4. **SUMMARIZE**

5. **SUPPRESS.**

The boldface numbers in the output show the effect of each option on the first break in the report. The PAGE option causes the report for each department to appear on a separate page.

BREAK *continued*

Output 9.4
Visual Affect of
Some of the
Options in the
BREAK Window

```
              Selecting DOUBLE OVERLINE, DOUBLE UNDERLINE, PAGE, SUMMARIZE         1
                       and SUPPRESS in the BREAK Window

              DEPT        ACCOUNT        BUDGET
              Equipment   lease       $80,000.00
                          maint       $22,000.00
                          purchase    $60,000.00
                          rental      $10,000.00
                          sets        $15,000.00
                          tape        $20,000.00
                                      ===========   1
                 5              4     $207,000.00
                                      ===========   2
```

```
              Selecting DOUBLE OVERLINE, DOUBLE UNDERLINE, PAGE, SUMMARIZE  3      2
                       and SUPPRESS in the BREAK Window

              DEPT        ACCOUNT        BUDGET
              Facilities  rent        $48,000.00
                          supplies     $5,500.00
                          utils        $8,500.00
                                      ===========
                                      $62,000.00
                                      ===========
```

```
              Selecting DOUBLE OVERLINE, DOUBLE UNDERLINE, PAGE, SUMMARIZE         3
                       and SUPPRESS in the BREAK Window

              DEPT        ACCOUNT        BUDGET
              Other       advert      $60,000.00
                          musicfee     $8,000.00
                          talent      $33,000.00
                                      ===========
                                      $101,000.00
                                      ===========
```

```
              Selecting DOUBLE OVERLINE, DOUBLE UNDERLINE, PAGE, SUMMARIZE         4
                       and SUPPRESS in the BREAK Window

              DEPT        ACCOUNT        BUDGET
              Staff       fulltime   $295,000.00
                          parttime   $100,000.00
                                      ===========
                                      $395,000.00
                                      ===========
```

```
              Selecting DOUBLE OVERLINE, DOUBLE UNDERLINE, PAGE, SUMMARIZE         5
                       and SUPPRESS in the BREAK Window

              DEPT        ACCOUNT        BUDGET
              Travel      gas          $2,000.00
                          leases       $8,000.00
                                      ===========
                                      $10,000.00
                                      ===========
```

COMPUTE

Uses a subset of DATA step statements and the LINE statement to define computed variables, perform calculations, and customize break lines

Display

```
┌REPORT─────────────────────────────────────────────────────────
│File Edit Search View Locals Globals
│
│                            The SAS System                      1
│
│                  ┌BREAK─────────────────────────────
│                  │       Breaking AFTER  DEPT
│                  │Options                      Color
┌COMPUTE ──────────────────────────────────────────────────────
│File View Globals
│
│00001
│00002
│00003
│00004
│00005
│00006
│00007
│00008
│00009
│00010
│00011
│00012
│                                                              R┘
```

Access

Select the **Edit Program** push button from the BREAK, DEFINITION, or COMPUTED VAR window.

Description

The COMPUTE window attaches DATA step statements, LINE statements, or both to part of the report. Use the SAS text editor commands to manipulate text in this window. For information on the text editor commands, see Chapter 19, "SAS Text Editor Commands," in *SAS Language: Reference, Version 6, First Edition.*

The DATA step statements do one of the following:

□ define a computed variable in the report (from either the COMPUTED VAR or DEFINITION window)

□ calculate values for a set of detail rows to use in break lines (from the BREAK window)

□ define break lines that contain text, calculated values, or both (from the BREAK window).

COMPUTE *continued*

When you attach DATA step statements to a computed variable, the REPORT procedure executes these statements once for each detail row and each break of the report. The value of a computed variable that appears in any given row of a report is the last value assigned to that variable's name during that execution of the DATA step statements. Because the REPORT procedure assigns values to the columns in a row of a report from left to right, a computed variable can depend only on values to its left.

When you attach DATA step statements to a break, the REPORT procedure executes these statements once at each break. Because PROC REPORT calculates statistics for groups before it actually constructs the rows of the report, statistics for sets of detail rows are available before or after the rows are displayed, as are values for any variables based on these statistics. For instance, if you want to display the mean of the variable BUDGET, add MEAN above or below BUDGET, or specify **MEAN** in the Statistic field in the DEFINITION window for BUDGET. The mean of BUDGET for all observations that have a unique set of values for all group and order variables is then available in the variable BUDGET.MEAN at the break, whether the break is before or after the detail rows. You can use this value to calculate other values that appear in the summary line.

On the other hand, some values are not available until the REPORT procedure has constructed the last detail row of a group. For example, you may want to determine the maximum difference between BUDGET and ACTUAL within each department. In such a case, you must compute this difference for each detail row and determine in which row the difference is greatest. The REPORT procedure cannot know which difference is greatest until it has processed the entire set of detail rows. Therefore, if you want to use the maximum difference in any break lines, the break lines must appear after the detail rows.

Chapter 6 provides details on the process of building a report.

COMPUTED VAR

Prompts you for the name of the computed variable you are adding to the report

Display

```
┌REPORT─────────────────────────────────────────────────────────────────
│File Edit Search View Locals Globals
│
│                            The SAS System                              1
│
│               DEPT        ACCOUNT        BUDGET
│               Equipment   lease       $80,000.00
│                           maint       $22,000.00
│                           purchas┌COMPUTED VAR────────────────────┐
│                           rental  │                                │
│                           sets    │  Variable name:                │
│                           tape    │  _ Character data              │
│               Facilities  rent    │    Length =                    │
│                           supplie │                                │
│                           utils   │  <Edit Program>   <OK>   <Cancel>│
│               Other       advert  │                                │
│                           musicfe └────────────────────────────────┘
│                           talent      $33,000.00
│
┌RKEYS───────────────────────────────────────────────────────────────────
│Add_above <F1>      Add_below <F2>      Add_left  <F3>     Add_right <F4>
│Break     <F5>      CGrow     <F6>      CShrink   <F7>     Define    <F8>
│Delete    <F9>      Move   <SHF F0>     Page   <SHF F1>    Pageback <SHF F2>
│                                                                    R
```

Access

Select **Computed variable** from the ADDING window.

Description

The COMPUTED VAR window prompts you for the name of the computed variable you are adding to the report. Enter the name of the variable at the prompt. If the variable is a character variable, select the **Character data** check box and, if you want, enter a value for the **Length** attribute. The length can be any integer between 1 and 200. If you leave the field blank, the REPORT procedure assigns a length of 8 to the variable.

After you enter the name of the variable, select the **Edit Program** push button, which opens the COMPUTE window. Use DATA step statements in the COMPUTE window to define the computed variable.

Note: Initially, the REPORT procedure uses the name of a computed variable as the header for the column or columns containing the values of that variable, just as it does for data set variables. You can change the column header by typing over it, by adding more lines to the header with the ADD_ABOVE or ADD_BELOW command, or by altering the **Header** field in the DEFINITION window for the variable. You can also change other characteristics, such as format and width, in the DEFINITION window.

DATASET VARS

Lists the variables in the data set and enables you to select the ones you want to add to the report

Display

```
┌REPORT────────────────────────────────────────────────────────────────────┐
│File Edit Search View Locals Globals                                       │
│                                                                           │
│                            The SAS System                              1  │
│                                                                           │
│                   DEPT        ACCOUNT        BUDGET                        │
│                   Equipment   lease        $80,0┌DATASET VARS────────┐     │
│                               maint        $22,0│File View Globals   │     │
│                               purchase     $60,0│                    │     │
│                               rental       $10,0│QTR                 │     │
│                               sets         $15,0│DEPT                │     │
│                               tape         $20,0│ACCOUNT             │     │
│                   Facilities  rent         $48,0│BUDGET              │     │
│                               supplies      $5,5│ACTUAL              │     │
│                               utils         $8,5└────────────────────┘     │
│                   Other       advert       $60,0                           │
│                               musicfee     $8,000.00                       │
│                               talent       $33,000.00                      │
│                                                                           │
├RKEYS──────────────────────────────────────────────────────────────────────┤
│Add_above  <F1>      Add_below  <F2>      Add_left  <F3>      Add_right <F4>│
│Break      <F5>      CGrow      <F6>      CShrink   <F7>      Define    <F8>│
│Delete     <F9>      Move   <SHF F0>      Page  <SHF F1>    Pageback <SHF F2>│
└──────────────────────────────────────────────────────────────────────R────┘
```

Access

□   Select **Dataset variable** from the ADDING window.

□   Invoke PROC REPORT with the PROMPT option. (The DATASET VARS window is one of the windows initially displayed.)

Description

The DATASET VARS window lists all variables in the data set. Select one or more variables to add to the report. When you select the first variable, it moves to the top of the list in the DATASET VARS window. If you select multiple variables, subsequent selections move to the bottom of the list of selected variables. The order of selected variables from top to bottom determines their order in the report from left to right.

DEFINITION

Displays the characteristics associated with an item in the report and enables you to change them

Display

```
┌REPORT─────────────────────────────────────────────────────────
 File Edit Search View Locals Globals

                          The SAS System                          1

      ┌DEFINITION────────────────────────────────────────────
      │            Definition of BUDGET
      │ Usage      Attributes              Options       Color
      │ o DISPLAY  Format    = DOLLAR11.2  _ NOPRINT     BLUE
      │ o ORDER    Spacing   = 2           _ NOZERO      RED
      │ o GROUP    Width     = 11          _ DESCENDING  PINK
      │ o ACROSS   Item help =             _ PAGE        GREEN
      │ * ANALYSIS Statistic = SUM                       CYAN
      │ o COMPUTED                                       YELLOW
      │            Type of data        Justification     WHITE
      │              * Numeric           o LEFT          ORANGE
      │              o Character         * RIGHT         BLACK
      │                                  o CENTER        MAGENTA
      │                                                  GRAY
 ┌RKEYS─────         │ Header = BUDGET                              BROWN
 Add_above <F1│
 Break     <F5│         <Edit Program>    <OK>     <Cancel>
 Delete    <F9│                                              ─R─
```

Access

□ Add a variable to the report from the DATASET VAR window.

□ Issue the DEFINE command.

Description

The DEFINITION window defines characteristics for an item in the report. The window contains seven different kinds of characteristics: **Usage, Attributes, Options, Type of data, Justification, Color,** and **Header.**

Usage
PROC REPORT uses a variable in one of six ways: as a display variable, as an order variable, as a group variable, as an across variable, as an analysis variable, or as a computed variable. The first five types of usage apply to data set variables; the sixth applies to variables you compute for the report.

   The way you use the variables in a report determines, among other things, the order of the detail rows in the report. The descriptions below explain the impact of each type of usage on the order of detail rows in the report. The next section, "Interactions of position and usage," explains how PROC REPORT orders detail rows when the report includes variables with a variety of usages.

DEFINITION *continued*

display variables

do not affect the order of the rows in the report. If the report contains no order or group variables to the left of a display variable, the order of detail rows in the report is the same as the order of observations in the data set. A report that contains one or more display variables has a detail row for each observation in the data set. Each detail row contains a value for each display variable. By default, the REPORT procedure treats all character variables as display variables. In Output 9.5 all variables are display variables.

Output 9.5
Displaying Display Variables

```
                     All Variables as Display Variables                    1

         QTR  DEPT        ACCOUNT       BUDGET        ACTUAL
          1   Staff       fulltime   $130,000.00   $127,642.68
          2   Staff       fulltime   $165,000.00   $166,345.75
          1   Staff       parttime    $40,000.00    $43,850.12
          2   Staff       parttime    $60,000.00    $56,018.96
          1   Equipment   lease       $40,000.00    $40,000.00
          2   Equipment   lease       $40,000.00    $40,000.00
          1   Equipment   purchase    $40,000.00    $48,282.38
          2   Equipment   purchase    $20,000.00    $17,769.15
          1   Equipment   tape         $8,000.00     $6,829.42
          2   Equipment   tape        $12,000.00    $11,426.73
          1   Equipment   sets         $7,500.00     $8,342.68
          2   Equipment   sets         $7,500.00     $8,079.62
          1   Equipment   maint       $10,000.00     $7,542.13
          2   Equipment   maint       $12,000.00    $10,675.29
          1   Equipment   rental       $4,000.00     $3,998.87
          2   Equipment   rental       $6,000.00     $5,482.94
          1   Facilities  rent        $24,000.00    $24,000.00
          2   Facilities  rent        $24,000.00    $24,000.00
          1   Facilities  utils        $5,000.00     $4,223.29
          2   Facilities  utils        $3,500.00     $3,444.81
          1   Facilities  supplies     $2,750.00     $2,216.55
          2   Facilities  supplies     $2,750.00     $2,742.48
          1   Travel      leases       $3,500.00     $3,045.15
          2   Travel      leases       $4,500.00     $3,889.65
          1   Travel      gas            $800.00       $537.26
          2   Travel      gas          $1,200.00       $984.93
          1   Other       advert      $30,000.00    $32,476.98
          2   Other       advert      $30,000.00    $37,325.64
          1   Other       talent      $13,500.00    $12,986.73
          2   Other       talent      $19,500.00    $18,424.64
          1   Other       musicfee     $3,000.00     $2,550.50
          2   Other       musicfee     $5,000.00     $4,875.95
```

order variables

order the detail rows in a report according to their formatted values. By default, the order is ascending, but you can alter it with the **DESCENDING** option in the DEFINITION window. If the report contains multiple order variables, the REPORT procedure establishes the order of the detail rows by treating these variables from left to right in the report. If a report contains one or more order variables, it contains a detail row for each observation in the data set.

PROC REPORT displays only the first occurrence of each value of an order variable in a set of detail rows that have the same values for all order variables. For example, in Output 9.6, DEPT and ACCOUNT are both order variables. QTR and BUDGET are display variables. Each value of the order variables appears only once for each department although the report contains two detail rows for each combination of values of DEPT and ACCOUNT.

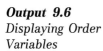

Output 9.6
Displaying Order
Variables

```
                  Using DEPT and ACCOUNT as Order Variables              1

          DEPT        ACCOUNT        QTR        BUDGET
          Equipment   lease          1       $40,000.00
                                     2       $40,000.00
                      maint          1       $10,000.00
                                     2       $12,000.00
                      purchase       1       $40,000.00
                                     2       $20,000.00
                      rental         1        $4,000.00
                                     2        $6,000.00
                      sets           1        $7,500.00
                                     2        $7,500.00
                      tape           1        $8,000.00
                                     2       $12,000.00
          Facilities  rent           1       $24,000.00
                                     2       $24,000.00
                      supplies       1        $2,750.00
                                     2        $2,750.00
                      utils          1        $5,000.00
                                     2        $3,500.00
          Other       advert         1       $30,000.00
                                     2       $30,000.00
                      musicfee       1        $3,000.00
                                     2        $5,000.00
                      talent         1       $13,500.00
                                     2       $19,500.00
          Staff       fulltime       1      $130,000.00
                                     2      $165,000.00
                      parttime       1       $40,000.00
                                     2       $60,000.00
          Travel      gas            1          $800.00
                                     2        $1,200.00
                      leases         1        $3,500.00
                                     2        $4,500.00
```

group variables
order the detail rows in a report according to their formatted values. By default, the order is ascending, but you can alter it with the **DESCENDING** option in the DEFINITION window. If a report contains one or more group variables, PROC REPORT tries to consolidate into one row all observations from the data set that have a unique combination of values for all group variables. The REPORT procedure cannot consolidate observations into groups if the report contains any order variables or any display variables that do not have one or more statistics above or below them. (PROC REPORT treats a display variable with statistics above or below it as an analysis variable.) If PROC REPORT cannot create groups, it returns a message and displays group variables as it displays order variables.

For example, in Output 9.7, DEPT and ACCOUNT are group variables. BUDGET is an analysis variable used to calculate the SUM statistic. Each row of this report shows the sum of the values for BUDGET for both quarters for one combination of DEPT and ACCOUNT.

Output 9.7
Creating Groups

```
                 Using DEPT and ACCOUNT as Group Variables               1

              DEPT        ACCOUNT        BUDGET
              Equipment   lease      $80,000.00
                          maint      $22,000.00
                          purchase   $60,000.00
                          rental     $10,000.00
                          sets       $15,000.00
                          tape       $20,000.00
              Facilities  rent       $48,000.00
                          supplies    $5,500.00
                          utils       $8,500.00
              Other       advert     $60,000.00
                          musicfee    $8,000.00
                          talent     $33,000.00
```

                                                    (continued on next page)

DEFINITION *(group variables continued)*

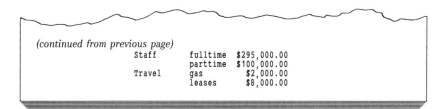

(continued from previous page)

```
         Staff    fulltime  $295,000.00
                  parttime  $100,000.00
         Travel   gas         $2,000.00
                  leases      $8,000.00
```

across variables
> are functionally similar to group variables; however, PROC REPORT displays the groups it creates for an across variable horizontally rather than vertically. Columns created by across variables contain statistical values. If the report does not have an analysis variable above or below an across variable, the values in the columns for the across variable contain a frequency count for the across variable.
>
> For example, in Output 9.8, QTR is a group variable and DEPT is an across variable. The report shows the number of observations in the data set for each department in each quarter.

Output 9.8
An Across
Variable without
an Analysis
Variable

```
       Displaying a Frequency Count for an Across Variable          1

                                  DEPT
   QTR  Equipment   Facilities   Other      Staff       Travel
    1        6           3         3          2            2
    2        6           3         3          2            2
```

Output 9.9 shows the same report with BUDGET added above DEPT as an analysis variable used to calculate the SUM statistic. In this report the values displayed in each column created for DEPT represent the sum of the budget amounts allocated to all accounts within that department for one quarter.

Output 9.9
An Across
Variable with an
Analysis Variable

```
       Displaying an Analysis Variable above an Across Variable      1

                                 BUDGET
                                  DEPT
   QTR    Equipment    Facilities     Other       Staff       Travel
    1    $109,500.00   $31,750.00  $46,500.00  $170,000.00   $4,300.00
    2     $97,500.00   $30,250.00  $54,500.00  $225,000.00   $5,700.00
```

analysis variables
> are used to calculate a statistic. You specify this statistic in the **Statistic** field of the DEFINITION window. You must specify a statistic for an analysis variable. By default, PROC REPORT uses numeric variables as analysis variables used to calculate the SUM statistic.
>
> The value displayed for an analysis variable is the value of the statistic you specify calculated for the set of observations represented by that row and column of the report. If the report contains one or more display or order variables, each detail row of the report represents one observation in the data set. However, if you create groups in the report, one detail row represents

multiple observations that have a unique combination of values for all the group variables.

If you create a break in the report on a group or order variable and elect to summarize in the break, the summary row includes all statistics and computed variables. In summary rows, the REPORT procedure calculates the statistic associated with an analysis variable over the entire set of observations that have the same value of the break variable.

For example, in Output 9.10, DEPT and ACCOUNT are group variables. BUDGET is an analysis variable used to calculate the MEAN statistic. Each detail row of the report shows the mean value of BUDGET for a particular combination of values for DEPT and ACCOUNT. For example, the first row shows that the mean value of BUDGET for the lease account in the equipment department is $40,000.00.

The report uses DEPT as a break variable, and the break lines include a summary line. Each summary line shows the mean of BUDGET for all accounts in one department.

Output 9.10
Using an Analysis Variable with Group Variables

```
Creating Summaries with Group and Analysis Variables          1

        DEPT        ACCOUNT      BUDGET
        Equipment   lease     $40,000.00
                    maint     $11,000.00
                    purchase  $30,000.00
                    rental     $5,000.00
                    sets       $7,500.00
                    tape      $10,000.00
        ----------            ----------
        Equipment             $17,250.00
        ----------            ----------

        Facilities  rent      $24,000.00
                    supplies   $2,750.00
                    utils      $4,250.00
        ----------            ----------
        Facilities            $10,333.33
        ----------            ----------

        Other       advert    $30,000.00
                    musicfee   $4,000.00
                    talent    $16,500.00
        ----------            ----------
        Other                 $16,833.33
        ----------            ----------

        Staff       fulltime $147,500.00
                    parttime  $50,000.00
        ----------            ----------
        Staff                 $98,750.00
        ----------            ----------

        Travel      gas        $1,000.00
                    leases     $4,000.00
        ----------            ----------
        Travel                 $2,500.00
        ----------            ----------
```

Now consider using an analysis variable with order variables. In Output 9.11, DEPT and ACCOUNT are order variables. BUDGET is an analysis variable used to calculate the SUM statistic. Each detail row of the report shows the sum of BUDGET for one observation in the report. (Because the report contains one detail row for each observation in the data set, the sum of BUDGET is the same as the value of BUDGET.) The structural difference between this report and the report in Output 9.10 is that this report does not consolidate observations for the first and second quarters that have a unique combination of values of DEPT and ACCOUNT and, therefore,

---

DEFINITION *(analysis variables continued)*

contains twice as many detail rows. Although this report doesn't consolidate observations, using BUDGET as an analysis variable enables you to use BUDGET in summary rows, as shown in Output 9.11. In this report each summary row shows the sum of BUDGET for all accounts in one department.

Output 9.11
Using an Analysis
Variable with
Order Variables

```
          Creating Summaries with Order and Analysis Variables        1

              DEPT       ACCOUNT       BUDGET
              Equipment  lease      $40,000.00
                                    $40,000.00
                         maint      $10,000.00
                                    $12,000.00
                         purchase   $40,000.00
                                    $20,000.00
                         rental      $4,000.00
                                     $6,000.00
                         sets        $7,500.00
                                     $7,500.00
                         tape        $8,000.00
                                    $12,000.00
              ----------            ----------
              Equipment           $207,000.00
              ----------            ----------

              Facilities rent      $24,000.00
                                   $24,000.00
                         supplies   $2,750.00
                                    $2,750.00
                         utils      $5,000.00
                                    $3,500.00
              ----------            ----------
              Facilities           $62,000.00
              ----------            ----------

              Other      advert    $30,000.00
                                   $30,000.00
                         musicfee   $3,000.00
                                    $5,000.00
                         talent    $13,500.00
                                   $19,500.00
              ----------            ----------
              Other               $101,000.00
              ----------            ----------

              Staff      fulltime $130,000.00
                                  $165,000.00
                         parttime  $40,000.00
                                   $60,000.00
              ----------            ----------
              Staff               $395,000.00
              ----------            ----------

              Travel     gas          $800.00
                                     $1,200.00
                         leases      $3,500.00
                                     $4,500.00
              ----------            ----------
              Travel               $10,000.00
              ----------            ----------
```

computed variables
are variables you define for the report. They are not in the data set. You add a computed variable to a report by selecting **Computed variable** from the ADDING window. When you add a computed variable, the REPORT procedure automatically sets its usage to computed in the DEFINITION window. You cannot change the usage of a computed variable.

Computed variables can be either numeric or character variables.

Interactions of position and usage   The position and usage of each variable in the report determine the report's structure and content. PROC REPORT orders the detail rows of the report according to the values of order and group variables, considered from left to right in the report.

For example, in Output 9.12 and Output 9.13, DEPT and QTR are both group variables and BUDGET is an analysis variable used to calculate the SUM statistic. The only difference in the reports is the relative positions of DEPT and QTR. The REPORT procedure groups first on the leftmost group variable.

Output 9.12
Grouping First by DEPT, Then by QTR

```
                    PROC REPORT Groups from Left to Right              1

                    DEPT         QTR       BUDGET
                    Equipment      1    $109,500.00
                                   2     $97,500.00
                    Facilities     1     $31,750.00
                                   2     $30,250.00
                    Other          1     $46,500.00
                                   2     $54,500.00
                    Staff          1    $170,000.00
                                   2    $225,000.00
                    Travel         1      $4,300.00
                                   2      $5,700.00
```

Output 9.13
Grouping First by QTR, Then by DEPT

```
                    PROC REPORT Groups from Left to Right              1

                    QTR   DEPT            BUDGET
                      1   Equipment   $109,500.00
                          Facilities   $31,750.00
                          Other        $46,500.00
                          Staff       $170,000.00
                          Travel        $4,300.00
                      2   Equipment    $97,500.00
                          Facilities   $30,250.00
                          Other        $54,500.00
                          Staff       $225,000.00
                          Travel        $5,700.00
```

Similarly, PROC REPORT orders columns of the report according to the values of across variables, considered from top to bottom. The difference in the appearance of the reports in Output 9.14 and Output 9.15 is due only to the difference in position of QTR and DEPT. In both cases QTR and DEPT are across variables and BUDGET is an analysis variable used to calculate the SUM statistic. In Output 9.14 QTR appears above DEPT, whereas in Output 9.15 DEPT appears above QTR.

Output 9.14
Grouping (across) First by QTR, Then by DEPT

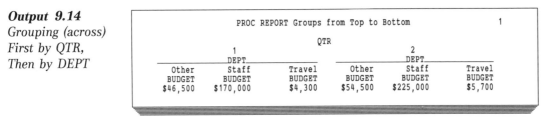

DEFINITION *continued*

Output 9.15
Grouping (across)
First by DEPT,
Then by QTR

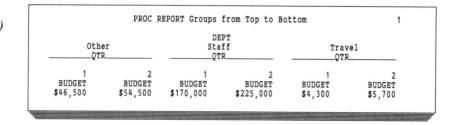

```
                 PROC REPORT Groups from Top to Bottom                1

                                   DEPT
            Other                  Staff                 Travel
        ____QTR____              ____QTR____           ____QTR____

          1          2            1          2           1          2
       BUDGET     BUDGET       BUDGET     BUDGET      BUDGET     BUDGET
       $46,500    $54,500     $170,000   $225,000     $4,300     $5,700
```

As you can see in Output 9.14 and Output 9.15, several items can collectively define the contents of a column in a report. For instance, in Output 9.15, the value for BUDGET that appears in each column is collectively determined by the variables BUDGET, QTR, and DEPT. Each value represents the sum of values of BUDGET for a particular department and a particular quarter. When you use multiple items to define the contents of a column, you can have at most one of the following in a column:

□ a display variable with or without a statistic above or below it

□ an analysis variable with or without a statistic above or below it

□ an order variable

□ a group variable

□ a computed variable

□ the N statistic.

Having more than one of these items in a column creates a conflict for the REPORT procedure about which values to display.

Table 9.1 shows with which other items each type of report item can share a column.

Note: Group and order variables cannot share a column with any other item.

Table 9.1   *Sharing Columns in a Report*

	Display	**Analysis**	**Order**	**Group**	**Computed**	**Across**	**Statistic**
Display	No	No	No	No	No	Yes	Yes
Analysis	No	No	No	No	No	Yes	Yes
Order	No	No	No	No	No	No	No
Group	No	No	No	No	No	No	No
Computed variable	No	No	No	No	No	Yes	No
Across variable	Yes	Yes	No	No	Yes	Yes	Yes
Statistic	Yes	Yes	No	No	No	Yes	No

The implications of the information in this table for adding items to the report are described below:

□ display variables

  □ You can add statistics above or below a numeric display variable. The report procedure treats each column as an analysis variable with the added statistic.

  □ You can add an across variable above or below a display variable.

□ analysis variables

  □ You can add statistics above or below an analysis variable. If you do so, the REPORT procedure does not calculate the statistic associated with the analysis variable in the DEFINITION window. Instead, it calculates the statistics you add.

  □ You can add an across variable above or below an analysis variable.

□ across variables

  □ You can add multiple across variables above or below an across variable. Multiple across variables do not cause a conflict because the REPORT procedure displays each value for an across variable only once, at the top of a column.

  □ You can add a single analysis variable or a single display variable above or below an across variable. Statistics can share a column with an across variable if an analysis variable or a display variable also shares that column.

□ computed variables

  □ You can add an across variable above or below a computed variable.

□ The following items can stand alone in a column:

  display variable

  analysis variable

  order variable

  group variable

  computed variable

  across variable

  N statistic.

  Note: The values in a column occupied only by an across variable are frequency counts.

DEFINITION *continued*

Attributes

The DEFINITION window defines five attributes for each item in the report. To alter the values of these attributes, simply type the value you want in the appropriate field. The five attributes are described below:

Format

> defines the format for the selected item. This format applies to the variable as PROC REPORT displays it; the format does not alter the format stored in the data set. By default, PROC REPORT uses the format you specify in the FORMAT statement when you invoke the REPORT procedure. If you do not specify a format, PROC REPORT uses the format stored in the data set. If no format for the item is stored in the data set, it uses the default format of BEST*w*. for numeric variables and $8. for character variables. PROC REPORT uses the default column width (the value of the COLWIDTH= option) as the width for the BEST*w*. format.
>
> If you do not know what format to use, enter a question mark (?) in the **Format** field and press ENTER. PROC REPORT opens the FORMAT window, which displays several commonly used SAS formats. For further information, see the documentation for the FORMAT window later in this chapter.
>
> **Note:** Changing the format of a group, order, or across variable can alter the order in which PROC REPORT displays the items in a report.

Spacing

> defines the number of horizontal positions to leave blank between the selected column and the column immediately to its left, if there is one.
>
> **Note:** The spacing between columns generated by an across variable with nothing below it is determined by the value of the spacing attribute in the ROPTIONS window.

Width

> defines the width of the column in which PROC REPORT displays the selected item. The value for the **Width** attribute can range from 1 to the value of the Linesize= option in the ROPTIONS window.
>
> If the definitions for items that share a column specify different widths, PROC REPORT uses the width of the item closest to the first row of the report. For example, in the report in Output 9.16 'Quarter' and 'Amount Budgeted' share a column. The width of 'Quarter' is 7, whereas the width of 'Amount Budgeted' is 11. Because 'Amount Budgeted' is closer to the first row of the report, the column that these two variables share has a width of 11.

Output 9.16
Conflicting Widths
in the Same
Column

```
                The Column Shared by Quarter and Amount Budgeted Takes       1
                   Its Width from the Definition for Amount Budgeted

                                        Quarter
                                      1              2
                                   Amount         Amount
                    Department     Budgeted       Budgeted
                    Equipment      $109,500.00    $97,500.00
                    Facilities     $31,750.00     $30,250.00
                    Other          $46,500.00     $54,500.00
                    Staff          $170,000.00    $225,000.00
                    Travel         $4,300.00      $5,700.00
```

However, if you use the same characteristics for 'Quarter' and 'Amount Budgeted' but place 'Quarter' below 'Amount Budgeted', the column the two variables share takes its width from the definition for 'Quarter'. Consequently, as Output 9.17 shows, the column is not wide enough to accommodate the DOLLAR11.2 format for 'Amount Budgeted'.

Output 9.17
Conflicting Widths
in the Same
Column

```
                The Column Shared by Quarter and Amount Budgeted Takes       1
                     Its Width from the Definition for Quarter

                                       Amount
                                       Budgeted
                                       Quarter
                    Department         1          2
                    Equipment          109500     97500.0
                    Facilities         31750.0    30250.0
                    Other              46500.0    54500.0
                    Staff              170000     225000
                    Travel             4300.00    5700.00
```

Item help

refers to a user-defined CBT or HELP entry supplying help for the selected item. Use the BUILD procedure in SAS/AF software to create a CBT or HELP entry for a report item. (For information on the BUILD procedure, see Chapter 10, "The BUILD Procedure," in *SAS/AF Software: Usage and Reference, Version 6, First Edition*.) All HELP entries for a report must be in the same catalog, and you must specify that catalog in the ROPTIONS window. To open a help frame from the report, select an item and issue the HELP command. PROC REPORT looks first for a CBT entry with the name you specify for the **Item help** attribute in the DEFINITION window and displays the help frame that entry describes. If no such entry exists, it looks for a HELP entry with that name. If neither a CBT nor a HELP entry for the selected item exists, the REPORT procedure displays the help frame for PROC REPORT.

Statistic

associates a statistic with an analysis variable. You cannot use this field with any other kind of variable, and you must use this field with an analysis variable (see the description of analysis variables in "Usage" earlier in the documentation for the DEFINITION window).

Note: The REPORT procedure uses the name or label of the analysis variable as the default header for the column. You can add the name of the statistic to the column header just as you make any change to a header.

DEFINITION *(Statistic continued)*

A list of valid values for the **Statistic** attribute follows. PROC REPORT calculates all the statistics available through PROC MEANS except for SKEWNESS and KURTOSIS. For complete descriptions of these statistics, see the documentation for the STATISTICS window.

N	RANGE	T
NMISS	SUM	PRT
MEAN	USS	VAR
STD	CSS	SUMWGT
MIN	STDERR	
MAX	CV	

Options

You may select one or more of the following options in the DEFINITION window:

NOPRINT

suppresses the display of the selected item. Use this option if you do not want to show the item in the report but you need to use its values to calculate other values that you do use or to determine the order of the detail rows in the report. You can temporarily override this option by selecting the SHOWALL option from the ROPTIONS window.

NOZERO

suppresses the display of a column containing values of an analysis or computed variable if all values are zero or missing. You can temporarily override this option by selecting the SHOWALL option from the ROPTIONS window.

DESCENDING

reverses the order in which PROC REPORT displays a group, order, or across variable. For instance, when DEPT is the leftmost order variable, the REPORT procedure displays the rows in ascending (alphabetic) order of the value of DEPT. When the characteristics for DEPT include **DESCENDING**, the values appear in reverse alphabetic order.

    Note: PROC REPORT orders group, order, and across variables according to their formatted values.

PAGE

inserts a page break in the report just before the column containing the selected variable. For example, by default in the report in Output 9.18, the REPORT procedure inserts a page break when it can fit no more columns on a page, creating a report with seven columns on the first page and one column on the second page.

Output 9.18
Default Page Break

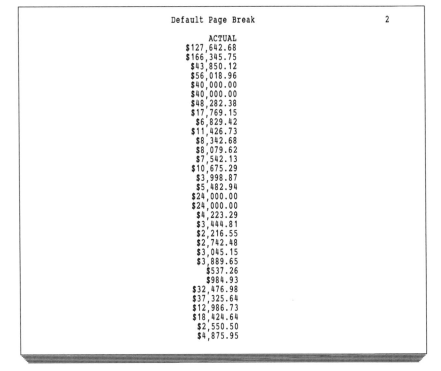

If you prefer to have the page break occur before the second occurrence of QTR, simply select the **PAGE** option in the DEFINITION window for that occurrence of QTR. The resulting report appears in Output 9.19.

DEFINITION *(PAGE continued)*

Output 9.19
Using the PAGE
Option to Control
the Position of the
Page Break

```
                                                                          1
              Controlling the Page Break with the PAGE Option
                        in the DEFINITION Window

              QTR  DEPT        ACCOUNT        BUDGET
               1   Staff       fulltime    $130,000.00
               2   Staff       fulltime    $165,000.00
               1   Staff       parttime     $40,000.00
               2   Staff       parttime     $60,000.00
               1   Equipment   lease        $40,000.00
               2   Equipment   lease        $40,000.00
               1   Equipment   purchase     $40,000.00
               2   Equipment   purchase     $20,000.00
               1   Equipment   tape          $8,000.00
               2   Equipment   tape         $12,000.00
               1   Equipment   sets          $7,500.00
               2   Equipment   sets          $7,500.00
               1   Equipment   maint        $10,000.00
               2   Equipment   maint        $12,000.00
               1   Equipment   rental        $4,000.00
               2   Equipment   rental        $6,000.00
               1   Facilities  rent         $24,000.00
               2   Facilities  rent         $24,000.00
               1   Facilities  utils         $5,000.00
               2   Facilities  utils         $3,500.00
               1   Facilities  supplies      $2,750.00
               2   Facilities  supplies      $2,750.00
               1   Travel      leases        $3,500.00
               2   Travel      leases        $4,500.00
               1   Travel      gas             $800.00
               2   Travel      gas           $1,200.00
               1   Other       advert       $30,000.00
               2   Other       advert       $30,000.00
               1   Other       talent       $13,500.00
               2   Other       talent       $19,500.00
               1   Other       musicfee      $3,000.00
               2   Other       musicfee      $5,000.00
```

```
                                                                          2
              Controlling the Page Break with the PAGE Option
                        in the DEFINITION Window

              QTR  DEPT        ACCOUNT        ACTUAL
               1   Staff       fulltime    $127,642.68
               2   Staff       fulltime    $166,345.75
               1   Staff       parttime     $43,850.12
               2   Staff       parttime     $56,018.96
               1   Equipment   lease        $40,000.00
               2   Equipment   lease        $40,000.00
               1   Equipment   purchase     $48,282.38
               2   Equipment   purchase     $17,769.15
               1   Equipment   tape          $6,829.42
               2   Equipment   tape         $11,426.73
               1   Equipment   sets          $8,342.68
               2   Equipment   sets          $8,079.62
               1   Equipment   maint         $7,542.13
               2   Equipment   maint        $10,675.29
               1   Equipment   rental        $3,998.87
               2   Equipment   rental        $5,482.94
               1   Facilities  rent         $24,000.00
               2   Facilities  rent         $24,000.00
               1   Facilities  utils         $4,223.29
               2   Facilities  utils         $3,444.81
               1   Facilities  supplies      $2,216.55
               2   Facilities  supplies      $2,742.48
               1   Travel      leases        $3,045.15
               2   Travel      leases        $3,889.65
               1   Travel      gas             $537.26
               2   Travel      gas             $984.93
               1   Other       advert       $32,476.98
               2   Other       advert       $37,325.64
               1   Other       talent       $12,986.73
               2   Other       talent       $18,424.64
               1   Other       musicfee      $2,550.50
               2   Other       musicfee      $4,875.95
```

Color

The color attribute controls the color of the column header and the values within the column. Available colors are

BLUE	**CYAN**	**BLACK**
RED	**YELLOW**	**MAGENTA**
PINK	**WHITE**	**GRAY**
GREEN	**ORANGE**	**BROWN**

Note: Not all operating systems and devices support all colors, and on some operating systems and devices, one color may map to another color. For example, if the DEFINITION window displays **BROWN** in yellow characters, selecting **BROWN** results in yellow values and a yellow column header.

Currently, color appears only when PROC REPORT displays the report in the REPORT window.

Type of Data

The SAS System uses a variable in one of two ways: as a character variable or as a numeric variable. The DEFINITION window for a report item indicates the item's type. You cannot change the type of an item in the DEFINITION window unless it is a computed variable.

Justification

You can justify the placement of the column header and of the values of an item within a column in one of three ways:

LEFT

left-justifies the formatted values of the specified item within the column width and left-justifies the column header over the values. If the format width is the same as the width of the column, LEFT has no effect.

RIGHT

right-justifies the formatted values of the specified item within the column width and right-justifies the column header over the values. If the format width is the same as the width of the column, RIGHT has no effect.

CENTER

centers the formatted values of the specified item within the column width and centers the column header over the values.

Note: When justifying values, PROC REPORT justifies the field width defined by the format of the item within the column. Thus, numbers are always aligned.

Header

You can enter up to 40 characters in the **Header** field. PROC REPORT uses the characters as the column header for the selected item. You can include the split character to split the header over multiple lines. The default split character is the slash (/). You can change the split character by altering the **Split** attribute in the ROPTIONS window.

If the first and last characters of a header are —, =, _, ., *, or +, or if the first and last characters of a header are < and > or > and <, the REPORT procedure expands the header by repeating the first and last characters to fill the space over the column.

FORMAT

Displays several commonly used SAS formats for you to choose from

Display

```
┌REPORT─────────────────────────────────────────────────────────────────
│File Edit Search View Locals Globals
│  ┌FORMATS─────────────────────────────────────────────────────┐      1
│  │                                                              │
│  │  Sample:    12345.68                                         │
│  │                                                              │
│  │  Width (w) =  11   Decimals (d) =   2              Color     │
│  │                                                    BLUE      │
│  │                 Numeric formats                    RED       │
│  │  * w.d          o BESTw.      o COMMAw.d   o DOLLARw.d  NG PINK│
│  │  o Ew.          o FRACTw.     o ROMANw.    o SSNw.     GREEN  │
│  │  o WORDFw.      o WORDSw.     o Zw.d                   CYAN   │
│  │                                                       YELLOW │
│  │          Date, time, and datetime formats             WHITE  │
│  │      o DATEw.           o DATETIMEw.d  o DDMMYYw.      ORANGE │
│  │      o HHMMw.d          o HOURw.d      o MMDDYYw.      BLACK  │
│  │                                                       MAGENTA│
│  │              <OK>            <Cancel>                  GRAY   │
│ RK                                                        BROWN  │
│ Ad└──────────────────────────────────────────────────────────────┘
│Break     <F5|     <Edit Program>    <OK>     <Cancel>
│Delete    <F9|                                                    R
```

Access

Enter a question mark (?) in the **Format** field in the DEFINITION window or in the PROMPTER window that prompts for format.

Description

The FORMAT window displays samples of a variety of numeric SAS formats. If you are uncertain about what format to use for a numeric variable in your report, use the FORMAT window to experiment with different formats. By changing the values for **Width** and **Decimals** and selecting one of the formats, you can see how a sample number looks with any width and decimal specification in any of the formats listed in the window. When you select **OK** to exit the FORMAT window, PROC REPORT assigns the **Width, Decimals,** and format you have selected to the variable in the DEFINITION window.

For further information on the use of the formats listed in the FORMAT window and on all SAS formats, refer to Chapter 14, "SAS Formats," in *SAS Language: Reference.*

LOCATION

Displays the choices for the location of break lines

Display

```
┌REPORT─────────────────────────────────────────────────────────────
 File Edit Search View Locals Globals

                          The SAS System                           1

                 DEPT       ACCOUNT       BUDGET
                 Equipment  lease      $80,000.00
                            maint      $22,000.00
                 ┌LOCATION─────────────┐  $60,000.00
                 │  o After detail     │  $10,000.00
                 │  o Before detail    │  $15,000.00
                 │                     │  $20,000.00
                 │  <OK>      <Cancel> │  $48,000.00
                 └─────────────────────┘   $5,500.00
                                           $8,500.00
                 Other      advert     $60,000.00
                            musicfee    $8,000.00
                            talent     $33,000.00

┌RKEYS──────────────────────────────────────────────────────────────
 Add_above <F1>      Add_below <F2>      Add_left  <F3>     Add_right <F4>
 Break     <F5>      CGrow     <F6>      CShrink   <F7>     Define    <F8>
 Delete    <F9>      Move    <SHF F0>    Page    <SHF F1>   Pageback <SHF F2>
                                                              ─R──
```

Access

Issue one of the following commands:

□ BREAK command

□ RBREAK command.

Description

Your choice in the LOCATION window determines whether break lines appear before or after the parts of the report to which they refer. A *break line* is a line of a report that contains one of the following:

□ characters that visually separate parts of the report.

□ summaries of statistics and computed variables (called a *summary line*).

□ text, values calculated for a set of rows of the report, or both. Chapter 5 and Chapter 6 explain how to use this kind of break line.

Use the BREAK window to perform calculations at a break or to define the contents of break lines. (The **OK** push button in the LOCATION window opens the BREAK window.)

The LOCATION window offers two choices: **After detail** and **Before detail**. The location of break lines depends on your choice in the LOCATION window and

---

LOCATION *continued*

whether you opened the LOCATION window with the BREAK command or the RBREAK command. Table 9.2 shows where the break lines are located.

Table 9.2
Locating Break
Lines in a Report

	After Detail	**Before Detail**
BREAK command	After the last detail row of each set of detail rows that have the same value for the selected group or order variable	Before the first detail row of each set of detail rows that have the same value for the selected group or order variable
RBREAK command	After the last row of the report	Before the first row of the report

For information on the importance of selecting **After detail** in certain situations, see "Usage" in the documentation of the BREAK command.

MESSAGE

Displays notes, warnings, and errors returned by the REPORT procedure, as well as the REPORT language generated by the LIST command

---

Display

```
┌MESSAGES──────────────────────────────────────────────
│File View Globals
│
│ ERROR: Statistic name is blank.
└──────────────────────────────────────────────────────
                  ┌DEFINITION────────────────────────────────────────
                  │              Definition of BUDGET
                  │  Usage      Attributes              Options        Color
                  │  o DISPLAY  Format    = DOLLAR11.2   _ NOPRINT     BLUE
                  │  o ORDER    Spacing   = 2            _ NOZERO      RED
                  │  o GROUP    Width     = 11           _ DESCENDING  PINK
                  │  o ACROSS   Item help =              _ PAGE        GREEN
                  │  * ANALYSIS Statistic =                            CYAN
                  │  o COMPUTED                                        YELLOW
                  │            Type of data        Justification      WHITE
                  │              * Numeric           o LEFT           ORANGE
                  │              o Character         * RIGHT          BLACK
                  │                                  o CENTER         MAGENTA
                  │                                                   GRAY
┌RKEYS──────┐    │  Header = BUDGET                                   BROWN
│Add_above <F1│  │
│Break     <F5│  │       <Edit Program>    <OK>     <Cancel>
│Delete    <F9│  │
                 └──────────────────────────────────────────────R────
```

Access

Issue the MESSAGE command.

Description

The MESSAGE window displays notes, warnings, and errors returned by the REPORT procedure, as well as the REPORT language generated by the LIST command. PROC REPORT clears all messages when you do one of the following:

□ open the DEFINITION window

□ add or delete an item

□ change information in the BREAK window

□ issue the LIST command

□ issue the END command in the MESSAGE window.

PROFILE

Enables you to customize certain aspects of the REPORT procedure

Display

```
┌REPORT─────────────────────────────────────────────────────────────────
│File Edit Search ┌PROFILE──────────────────────────────────────┐
│                 │                                            1 │
│                 │   Menu library:                              │
│                 │   Menu catalog:                              │
│                 │   Report window menu:                        │
│                 │   Compute window menu:                       │
│                 │                                              │
│                 │  Report options                              │
│                 │   X WINDOWS                                  │
│                 │   _ PROMPT                                   │
│                 │   _ COMMAND                                  │
│                 │   _ NORKEYS                                  │
│                 │                                              │
│                 │     <Change display>  <OK>    <Cancel>       │
│                 │                                              │
│                 └──────────────────────────────────────────────┘
│                              talent     $33,000.00
┌RKEYS──────────────────────────────────────────────────────────────────
│Add_above <F1>      Add_below <F2>      Add_left  <F3>      Add_right <F4>
│Break     <F5>      CGrow     <F6>      CShrink   <F7>      Define    <F8>
│Delete    <F9>      Move      <SHF F0>  Page      <SHF F1>  Pageback  <SHF F2>
└──────────────────────────────────────────────────────────────R─────────
```

Access

Issue the PROFILE command.

PROFILE *continued*

Description

The PROFILE window enables you to

□ specify the location of menus that define alternative action bars and pull-down menus for the REPORT and COMPUTE windows

□ set defaults for selected options

□ change the colors of most REPORT windows.

When you use the **OK** push button to exit the PROFILE window, the REPORT procedure stores your profile in SASUSER.PROFILE.REPORT.PROFILE. If you cancel, rather than exit, the window, PROC REPORT does not save the profile.

Specifying Alternative Action Bars and Pull-Down Menus

You can use the PMENU procedure to create catalog entries of type MENU that create your own action bars and pull-down menus for the REPORT and COMPUTE windows. The first four fields in the PROFILE window enable you to use these action bars and pmenus.

Menu library
contains the libref for the SAS data library that contains the catalog in which the MENU entries you create with PROC PMENU reside.

Menu catalog
contains the name of the catalog that contains your MENU entries.

Report window menu
contains the name of the menu that creates the action bar and pull-down menus for the REPORT window.

Compute window menu
contains the name of the menu that creates the action bar and pull-down menus for the COMPUTE window.

Report Options

If you set an option in the PROFILE window, PROC REPORT uses that option whenever you invoke the procedure unless you specifically override the option in the PROC REPORT statement. For information on the function of the WINDOWS, PROMPT, COMMAND, and NORKEYS options, see Chapter 10, "The REPORT Language."

Changing Colors in Windows

In addition to the usual **OK** and **CANCEL** push buttons, the PROFILE window contains the **Change display** push button. If you select this push button, PROC REPORT displays the WINDOW SELECT window.

```
┌REPORT──────────────────────────────────────────────────────────────────
│File Edit Search ┌PROFILE──────────────────────────────────────
│                 │┌WINDOW SELECT─────────────────────────────┐
│                 ││    o ADDING                              │         1
│                 ││    o BREAK                               │
│                 ││    o COMPUTED VAR                        │
│                 ││    o DATASET VARS                        │
│                 ││    o DEFINITION                          │
│                 ││    o FORMATS                             │
│                 ││    o LOCATION                            │
│                 ││    o MESSAGE                             │
│                 ││    o PROFILE                             │
│                 ││    o PROMPTER                            │
│                 ││    o REPORT                  ncel>       │
│                 ││    o RKEYS                               │
│                 ││    o RLOAD                               │
│                 ││    o ROPTIONS                00.00       │
│                 ││    o RSTORE                              │
│                 ││    o STATISTICS                          │
┌RKEYS────────────│                                          │
│Add_above <F1>   │        <Colors>   <OK>          │<F3>   Add_right <F4>
│Break     <F5>   └──────────────────────────────────────────┘<F7>   Define    <F8>
│Delete    <F9>                                   <SHF F1> Pageback <SHF F2>
│                                                                            R
```

When you select a window from the WINDOW SELECT window and select the **Colors** push button, PROC REPORT displays the WINDOW COLORS window.

```
┌REPORT──────────────────────────────────────────────────────────────────
│File Edit Search ┌PROFILE──────────────────────────────────────
│                 │┌WINDOW SELECT─────────────────────────────┐
│┌WINDOW COLORS───────────────────────────────────────────────┐           1
││                                                             │
││    Back     Banner     Border    Command    Message    Text │
││    Color     Color      Color      Color      Color    Color│
││  BLUE       BLUE       BLUE       BLUE       BLUE       BLUE │
││  RED        RED        RED        RED        RED        RED  │
││  PINK       PINK       PINK       PINK       PINK       PINK │
││  GREEN      GREEN      GREEN      GREEN      GREEN      GREEN│
││  CYAN       CYAN       CYAN       CYAN       CYAN       CYAN │
││  YELLOW     YELLOW     YELLOW     YELLOW     YELLOW     YELLOW│
││  WHITE      WHITE      WHITE      WHITE      WHITE      WHITE│
││  ORANGE     ORANGE     ORANGE     ORANGE     ORANGE     ORANGE│
││  BLACK      BLACK      BLACK      BLACK      BLACK      BLACK│
││  MAGENTA    MAGENTA    MAGENTA    MAGENTA    MAGENTA    MAGENTA│
││  GRAY       GRAY       GRAY       GRAY       GRAY       GRAY │
││  BROWN      BROWN      BROWN      BROWN      BROWN      BROWN│
┌RK                                                            │
│Ad                       <OK>      <Cancel>                   │right <F4>
│Br                                                            │ne    <F8>
│De└─────────────────────────────────────────────────────────┘back  <SHF F2>
│                                                                            R
```

PROFILE *continued*

In this window, select the colors you want to use for the various parts of your selected window. Definitions of the parts of the window follow:

Back
is the background of the window. On some operating systems and devices, you cannot change the background color.

Banner
is the command prompt at the upper left corner of the window.

Border
is the lined border of the window and the window name.

Command
is the action bar or the unprotected part of the command line where you enter commands.

Message
is the message line immediately below the command line or action bar.

Text
is all text lines in the window, including future text to be typed in.

PROMPTER

Prompts you for information as you build a report

Display

```
┌REPORT─────────────────────────────────────────┐   ┌DATASET VARS──────────┐
│ File Edit Search View Locals Globals           │   │ File View Globals    │
│                                                │   │                      │
│                                                │   │ QTR                  │
│                                                │   │ DEPT                 │
│                                                │   │ ACCOUNT              │
│                                                │   │ BUDGET               │
│                                                │   │ ACTUAL               │
│                                                │   │                      │
│                                                │   │                      │
│                                                │   │                      │
│ ┌PROMPTER────────────────────────────────────┐│   └──────────────────────┘
│ │                                            ││
│ │ Choose items for report by selecting       ││
│ │ variables from                             ││
│ │ the DATASET VARS window.                    ││
│ │                                            ││   ─────────────────────
│ │ For help place the cursor over an item      ││      Add_right  <F4>
│ │ and press                                   ││      Define     <F8>
│ │ the HELP function key.                      ││  1>  Pageback   <SHF F2>
│ │                                            ││   ─────────────────────R─
│ └────────────────────────────────────────────┘│
└────────────────────────────────────────────────┘
```

Access

□ Invoke PROC REPORT with the PROMPT option. (The PROMPTER window is one of the windows initially displayed.)

□ Select the **PROMPT** option in the ROPTIONS window and issue one of the adding commands.

Description

The PROMPTER window is a tool that prompts you for information as you add either data set variables or statistics to a report. You choose to use this tool by specifying the PROMPT option when you invoke the REPORT procedure or by selecting the **PROMPT** option from the ROPTIONS window.

The PROMPTER window steps you through the most commonly used parts of the windows you would use to add either a data set variable or a statistic to the report. However, the PROMPTER window provides more guidance than the other windows provide. The title of the PROMPTER window changes to the name of the window you would use to perform a task if you were not using the prompting facility so that you can begin to associate the windows with their functions and so that you know what window to use if you later decide to change something. For details on the choices in the PROMPTER window, see the documentation for the ADDING, BREAK, DATASET VARS, DEFINITION, and STATS windows as well as for the ADD_ABOVE, ADD_BELOW, ADD_LEFT, and ADD_RIGHT commands.

PROMPTER *continued*

The PROMPTER window may contain three push buttons: **OK, Cancel,** and **Backup.** The **OK** push button functions in the usual way. The **Cancel** push buttons cancels prompting at that point. The **Backup** push button returns you to the previous PROMPTER window.

REPORT

Contains the text of the report

Display

```
┌REPORT─────────────────────────────────────────────────────────────────────┐
│File Edit Search View Locals Globals                                        │
│                                                                            │
│                            The SAS System                                 1│
│                                                                            │
│                    DEPT        ACCOUNT       BUDGET                         │
│                    Equipment   lease      $80,000.00                        │
│                                maint      $22,000.00                        │
│                                purchase   $60,000.00                        │
│                                rental     $10,000.00                        │
│                                sets       $15,000.00                        │
│                                tape       $20,000.00                        │
│                    Facilities  rent       $48,000.00                        │
│                                supplies    $5,500.00                        │
│                                utils       $8,500.00                        │
│                    Other       advert     $60,000.00                        │
│                                musicfee    $8,000.00                        │
│                                talent     $33,000.00                        │
│                                                                            │
└────────────────────────────────────────────────────────────────────────────┘
┌RKEYS───────────────────────────────────────────────────────────────────────┐
│Add_above <F1>      Add_below <F2>     Add_left  <F3>      Add_right <F4>     │
│Break     <F5>      CGrow     <F6>     CShrink   <F7>      Define    <F8>     │
│Delete    <F9>      Move   <SHF F0>    Page   <SHF F1>     Pageback <SHF F2>  │
└──────────────────────────────────────────────────────────────────────────R─┘
```

Access

Invoke PROC REPORT with the WINDOWS or PROMPT option. (The REPORT window is one of the windows initially displayed.)

Description

The REPORT window contains the text of the report, including detail rows, break lines, titles, and headers. Most of the report is protected, that is, you can't write to most of the report directly; you must use the REPORT procedure to change it. However, you can type over column headers.

By default, the REPORT window includes an action bar that contains six items: **File, Edit, Search, View, Locals,** and **Globals.** The following list shows which commands appear in the pull-down menu under each selection in the action bar. For information on individual REPORT commands, see Chapter 8, "REPORT Commands."

File

displays a pull-down menu of REPORT commands that enable you to manage report definitions or to exit the procedure:

RSTORE
RLOAD
QUIT

Edit

displays a pull-down menu of REPORT commands that enable you to manipulate the current report definition:

ADD_ABOVE	CGROW	RBREAK
ADD_BELOW	CSHRINK	REFRESH
ADD_LEFT	DEFINE	SPAN
ADD_RIGHT	DELETE	
BREAK	MOVE	

Search

displays a menu of commands that enable you to use the WHERE clause within your PROC REPORT session:

Where
Where also
Undo last where

For information on the WHERE clause, see the documentation for the WHERE statement in Chapter 9, "SAS Language Statements" in *SAS Language: Reference.*

View

displays a pull-down menu of commands that enable you to view parts of the report that are not currently in the REPORT window. In this pull-down menu, only the PAGE and PAGEBACK commands are REPORT commands. For information on the other commands, see Chapter 18, "SAS Display Manager Commands," in *SAS Language: Reference.*

PAGE	FORWARD	LEFT
PAGEBACK	BACKWARD	RIGHT

REPORT *continued*

Locals

displays a pull-down menu of commands that access miscellaneous REPORT services and information:

ROPTIONS	LIST	PROFILE
MESSAGE	RKEYS	

Globals

displays a pull-down menu of SAS display manager commands that you may want to issue frequently while you are using the REPORT procedure. For information on these commands, see Chapter 18 in *SAS Language: Reference.*

OPTIONS	KEYS
TITLE	COMMAND
FOOTNOTE	

RKEYS

Displays definitions for up to 12 function keys defined as REPORT commands

Display

```
┌REPORT─────────────────────────────────────────────────────────────────────
File Edit Search View Locals Globals
Screen image saved to SASUSER.SCRNDUMP.
                          The SAS System                                    1

              DEPT        ACCOUNT        BUDGET
              Equipment   lease       $80,000.00
                          maint       $22,000.00
                          purchase    $60,000.00
                          rental      $10,000.00
                          sets        $15,000.00
                          tape        $20,000.00
              Facilities  rent        $48,000.00
                          supplies     $5,500.00
                          utils        $8,500.00
              Other       advert      $60,000.00
                          musicfee     $8,000.00
                          talent      $33,000.00

┌RKEYS──────────────────────────────────────────────────────────────────────
Add_above <F1>     Add_below <F2>     Add_left  <F3>     Add_right <F4>
Break     <F5>     CGrow     <F6>     CShrink   <F7>     Define    <F8>
Delete    <F9>     Move      <SHF F0> Page      <SHF F1> Pageback  <SHF F2>
                                                                          ─R─
```

Access

□ Invoke PROC REPORT. (By default, the RKEYS window is one of the windows initially displayed.)

□ Issue the RKEYS command.

Description

The RKEYS window displays definitions for up to 12 function keys defined as REPORT commands. The REPORT procedure provides a set of default key definitions that you can alter through the KEYS window, which you open with the KEYS command. For information on changing the default function key definitions, see "Defining a Function Key" in Chapter 7, "Using the REPORT Procedure in a Windowing Environment."

You can issue a REPORT command by selecting the command or the name of the corresponding function key from the RKEYS window.

To suppress the display of the RKEYS window, specify the NORKEYS option in the PROC REPORT statement.

Note: You can define more than 12 function keys, but the RKEYS window can display only 12 definitions.

RLOAD

Displays the names of all report entries stored in the specified catalog and enables you to load the report you select

Display

```
┌REPORT─────┌RLOAD──────────────────────────────────────┐
│File Edit S│File View Globals                          │
│           │                                           │
│           │Libname:                                 1 │
│           │Catalog:                                   │
│           │                                           │
│           │                                           │
│           └───────────────────────────────────────────┘
│                        purchase   $60,000.00
│                        rental     $10,000.00
│                        sets       $15,000.00
│                        tape       $20,000.00
│              Facilities rent       $48,000.00
│                        supplies    $5,500.00
│                        utils       $8,500.00
│              Other     advert     $60,000.00
│                        musicfee    $8,000.00
│                        talent     $33,000.00
├RKEYS──────────────────────────────────────────────────────
│Add_above <F1>     Add_below <F2>     Add_left  <F3>     Add_right <F4>
│Break     <F5>     CGrow     <F6>     CShrink   <F7>     Define    <F8>
│Delete    <F9>     Move      <SHF F0> Page      <SHF F1> Pageback  <SHF F2>
└───────────────────────────────────────────────────────R──
```

RLOAD   *continued*

```
┌REPORT──────┌RLOAD──────────────────────────────────────┐
│File Edit S │File View Globals                           │
│            │                                            │
│            │Libname: SASUSER                           ↑│
│            │Catalog: TUTORS                             │
│            │                                            │
│            │CHAP2     tutorials: chapter 2              │
│            │CHAP3     tutorials: chapter 3              │
│            │CHAP4     tutorials: chapter 4              │
│            │CHAP5A    tutorials: chapter 5(a)           │
│            │CHAP5B    tutorials: chapter 5(b)           │
│            │CHAP6     tutorials: chapter 6              │
│            └────────────────────────────────────────────┘
│                         supplies    $5,500.00
│                         utils       $8,500.00
│                 Other   advert      $60,000.00
│                         musicfee    $8,000.00
│                         talent      $33,000.00
│
┌RKEYS──────────────────────────────────────────────────────
│Add_above <F1>      Add_below <F2>      Add_left  <F3>      Add_right <F4>
│Break     <F5>      CGrow     <F6>      CShrink   <F7>      Define    <F8>
│Delete    <F9>      Move     <SHF F0>   Page     <SHF F1>   Pageback <SHF F2>
└──────────────────────────────────────────────────────────R
```

Access

Issue the RLOAD command.

Description

Use the RLOAD window to load a report definition that the RSTORE command or the OUTREPT= option previously stored. The SAS System stores the report definition as a catalog entry of type REPT. When you load a report definition, the REPORT procedure applies the definition to the variables in the data set you are using. You can use a report definition with any data set that contains variables with the same names as the ones used in the definition.

Descriptions of the fields in the RLOAD window follow:

Libname
contains the libref of the SAS data library that contains the catalog in which the report definition resides.

Catalog
contains the name of the catalog that contains the report definition.

After you specify a catalog and press ENTER, the REPORT procedure displays all the entries of type REPT stored in that catalog. Select the entry you want to use. If the report you want is not stored in the catalog you have specified, simply change the libref and catalog name until you find the entry you seek.

ROPTIONS

Displays choices that control layout and display of the entire report and identifies the SAS data library and catalog containing CBT or HELP entries for items in the report

Display

```
-REPORT-
File Edit Se-ROPTIONS-
          Modes         Attributes
          _ DEFER       Linesize  =  78
          _ PROMPT      Pagesize  =  20
                        Colwidth  =   9
          Options       Spacing   =   2
          X CENTER      Split     =   /
          _ HEADLINE    Panels    =   1
          _ HEADSKIP    Panelspace =  4
          _ NAMED
          _ NOHEADER         User Help
          _ SHOWALL     Libname =
          _ WRAP        Catalog =

              <OK>        <Cancel>
```

Access

Issue the ROPTIONS command.

Description

In the ROPTIONS window you make choices that have an impact on the layout and display of the entire report.

Modes

The mode you select determines whether the REPORT procedure immediately refreshes your report when you change it or waits until you issue the REFRESH command. The two modes available are described here:

DEFER

stores the information for changes and makes them all at once when you issue the REFRESH command. By default, PROC REPORT redisplays the report in the REPORT window each time you redefine the report by adding or deleting an item, by changing information in the DEFINITION window, or by changing information in the BREAK window.

The DEFER mode is particularly useful when you know you need to make several changes to the report but do not want to see the intermediate reports.

ROPTIONS *continued*

PROMPT

turns on the REPORT procedure's prompting facility, which displays the PROMPTER window when you add either data set variables or statistics to the report. The PROMPTER window guides you through the addition of data set variables or statistics to your report.

Options

You can select one or more of the following options from the ROPTIONS window:

CENTER

centers the report within the specified line size. Selecting and canceling **CENTER** in the ROPTIONS window has no effect on the setting of the CENTER system option.

HEADLINE

underlines all column headers and the spaces between them at the top of each page of the report.

HEADSKIP

writes a blank line beneath all column headers (or beneath the underlining that the **HEADLINE** option writes) at the top of each page of the report.

NAMED

writes *name=* in front of each value in the report, where *name* is the column header for the value. You may find it useful to specify the **NAMED** option in conjunction with the **WRAP** option to produce a report that wraps all columns for a single row of the report onto consecutive lines rather than placing columns of a wide report on separate pages. When you select the **NAMED** option, PROC REPORT automatically uses the **NOHEADER** option. (Documentation for the **WRAP** and **NOHEADER** options appears later in this section.)

Output 9.20 contains a report in which the collective width of the columns exceeds the line size. By default, PROC REPORT displays values for as many columns of the report as it can fit on one page for all rows before displaying values for the remaining columns.

Output 9.20
Default Report

```
               Values for ACTUAL and BALANCE for the Second Quarter          1
                               Appear on Page 2

                                           QTR
                                            1                      2
        DEPT        ACCOUNT      BUDGET       ACTUAL    BALANCE      BUDGET
        Equipment   lease    $40,000.00   $40,000.00      $0.00  $40,000.00
                    maint    $10,000.00    $7,542.13  $-2457.87  $12,000.00
                    purchase $40,000.00   $48,282.38  $8,282.38  $20,000.00
                    rental    $4,000.00    $3,998.87     $-1.13   $6,000.00
                    sets      $7,500.00    $8,342.68    $842.68   $7,500.00
                    tape      $8,000.00    $6,829.42  $-1170.58  $12,000.00
        Facilities  rent     $24,000.00   $24,000.00      $0.00  $24,000.00
                    supplies  $2,750.00    $2,216.55   $-533.45   $2,750.00
                    utils     $5,000.00    $4,223.29   $-776.71   $3,500.00
```

```
    Other       advert      $30,000.00   $32,476.98    $2,476.98   $30,000.00
                musicfee     $3,000.00    $2,550.50     $-449.50    $5,000.00
                talent      $13,500.00   $12,986.73     $-513.27   $19,500.00
    Staff       fulltime   $130,000.00  $127,642.68    $-2357.32  $165,000.00
                parttime    $40,000.00   $43,850.12    $3,850.12   $60,000.00
    Travel      gas            $800.00      $537.26     $-262.74    $1,200.00
                leases       $3,500.00    $3,045.15     $-454.85    $4,500.00
```

```
           Values for ACTUAL and BALANCE for the Second Quarter         2
                          Appear on Page 2

                                  QTR
                                   2
                               ACTUAL      BALANCE
                              $40,000.00       $0.00
                              $10,675.29   $-1324.71
                              $17,769.15   $-2230.85
                               $5,482.94    $-517.06
                               $8,079.62     $579.62
                              $11,426.73    $-573.27
                              $24,000.00       $0.00
                               $2,742.48      $-7.52
                               $3,444.81     $-55.19
                              $37,325.64    $7,325.64
                               $4,875.95    $-124.05
                              $18,424.64   $-1075.36
                             $166,345.75    $1,345.75
                              $56,018.96   $-3981.04
                                 $984.93    $-215.07
                               $3,889.65    $-610.35
```

The same report appears in Output 9.21, but here the NAMED and WRAP options are in effect.

Output 9.21
Report with the WRAP and NAMED Options in Effect

```
           Report with the WRAP and NAMED Options: One Value from Each      1
                 Column Appears before Another Value from the First Column

DEPT=Equipment   ACCOUNT=lease      BUDGET= $40,000.00  ACTUAL= $40,000.00
BALANCE=     $0.00  BUDGET= $40,000.00  ACTUAL= $40,000.00  BALANCE=     $0.0
DEPT=            ACCOUNT=maint      BUDGET= $10,000.00  ACTUAL=  $7,542.13
BALANCE=$-2457.87  BUDGET= $12,000.00  ACTUAL= $10,675.29  BALANCE=$-1324.7
DEPT=            ACCOUNT=purchase   BUDGET= $40,000.00  ACTUAL= $48,282.38
BALANCE=$8,282.38  BUDGET= $20,000.00  ACTUAL= $17,769.15  BALANCE=$-2230.8
DEPT=            ACCOUNT=rental     BUDGET=  $4,000.00  ACTUAL=  $3,998.87
BALANCE=    $-1.13  BUDGET=  $6,000.00  ACTUAL=  $5,482.94  BALANCE= $-517.0
DEPT=            ACCOUNT=sets       BUDGET=  $7,500.00  ACTUAL=  $8,342.68
BALANCE=  $842.68  BUDGET=  $7,500.00  ACTUAL=  $8,079.62  BALANCE=  $579.6
DEPT=            ACCOUNT=tape       BUDGET=  $8,000.00  ACTUAL=  $6,829.42
BALANCE=$-1170.58  BUDGET= $12,000.00  ACTUAL= $11,426.73  BALANCE= $-573.2
DEPT=Facilities  ACCOUNT=rent       BUDGET= $24,000.00  ACTUAL= $24,000.00
BALANCE=     $0.00  BUDGET= $24,000.00  ACTUAL= $24,000.00  BALANCE=     $0.0
DEPT=            ACCOUNT=supplies   BUDGET=  $2,750.00  ACTUAL=  $2,216.55
BALANCE= $-533.45  BUDGET=  $2,750.00  ACTUAL=  $2,742.48  BALANCE=   $-7.5
DEPT=            ACCOUNT=utils      BUDGET=  $5,000.00  ACTUAL=  $4,223.29
BALANCE= $-776.71  BUDGET=  $3,500.00  ACTUAL=  $3,444.81  BALANCE=  $-55.1
DEPT=Other       ACCOUNT=advert     BUDGET= $30,000.00  ACTUAL= $32,476.98
BALANCE=$2,476.98  BUDGET= $30,000.00  ACTUAL= $37,325.64  BALANCE=$7,325.6
DEPT=            ACCOUNT=musicfee   BUDGET=  $3,000.00  ACTUAL=  $2,550.50
BALANCE= $-449.50  BUDGET=  $5,000.00  ACTUAL=  $4,875.95  BALANCE= $-124.0
DEPT=            ACCOUNT=talent     BUDGET= $13,500.00  ACTUAL= $12,986.73
BALANCE= $-513.27  BUDGET= $19,500.00  ACTUAL= $18,424.64  BALANCE=$-1075.3
DEPT=Staff       ACCOUNT=fulltime   BUDGET=$130,000.00  ACTUAL=$127,642.68
BALANCE=$-2357.32  BUDGET=$165,000.00  ACTUAL=$166,345.75  BALANCE=$1,345.7
DEPT=            ACCOUNT=parttime   BUDGET= $40,000.00  ACTUAL= $43,850.12
BALANCE=$3,850.12  BUDGET= $60,000.00  ACTUAL= $56,018.96  BALANCE=$-3981.0
DEPT=Travel      ACCOUNT=gas        BUDGET=    $800.00  ACTUAL=    $537.26
BALANCE= $-262.74  BUDGET=  $1,200.00  ACTUAL=    $984.93  BALANCE= $-215.0
DEPT=            ACCOUNT=leases     BUDGET=  $3,500.00  ACTUAL=  $3,045.15
BALANCE= $-454.85  BUDGET=  $4,500.00  ACTUAL=  $3,889.65  BALANCE= $-610.3
```

ROPTIONS *continued*

NOHEADER

suppresses column headers, including those that span multiple columns.

▶ *Caution* . . . . *Once you suppress the display of column headers, you cannot select any report items.* ▲

SHOWALL

overrides selections in the DEFINITION window that suppress the display of a column (**NOPRINT** and **NOZERO**).

WRAP

displays one value from each column of the report, on consecutive lines if necessary, before displaying another value from the first column. By default, PROC REPORT displays values for as many columns as it can fit on one page before it begins to display the remaining columns.

When you use the **WRAP** option, you may also want to use the **NAMED** option. An example of a report that uses the **WRAP** and **NAMED** options appears with the documentation for the **NAMED** option earlier in this section.

Attributes

The attributes in the ROPTIONS window determine the physical layout of the report. The procedure supplies a default value for each of these items. Change these values by typing over them. The attributes are described below:

Linesize

specifies the length of a line of the report. Valid values range from 64 to 256. The default value is the value of the LS= option in the PROC REPORT statement. If you don't specify the LS= option, the default value is the value stored with the report definition you are using. If you are not using a report definition, PROC REPORT takes its line size from the value of the LINESIZE= system option. Altering the line size in the ROPTIONS window has no effect on the value of the LINESIZE= system option.

If the value of the **Linesize** attribute is greater than the width of your REPORT window, use the SAS display manager commands RIGHT and LEFT to display portions of the report that are not currently in the display.

Pagesize

specifies the number of lines in a page of the report. Valid values range from 15 to 32,767. The default value is the value of the PS= option in the PROC REPORT statement. If you don't specify the PS= option, the default value is the value stored with the report definition you are using. If you are not using a report definition, PROC REPORT takes its page size from the value of the PAGESIZE= system option. Altering the page size in the ROPTIONS window has no effect on the value of the PAGESIZE= system option.

If the value of the **Pagesize** attribute is greater than the length of your REPORT window, use the SAS display manager commands FORWARD and BACKWARD to display portions of a page that are not currently in the display. If the report contains more than one page, use the REPORT commands PAGE and PAGEBACK to display different pages.

Colwidth

determines the default number of horizontal positions for all columns you add to the report. Valid values range from 1 to the line size. The default value is 9.

You can change the width of an individual column in the DEFINITION window for that column or with the CSHRINK or CGROW command.

Spacing

determines the default number of horizontal positions between any column you add to a report and the column immediately to its left, if there is one. The default value is 2.

Use the **Spacing** attribute in the DEFINITION window to change the number of horizontal positions that precede one particular column in the report. For each column, the sum of its width and the number of horizontal positions separating it from the column to its left cannot exceed the line size.

Split

specifies the split character. If you use the split character in a header, the REPORT procedure breaks the header when it reaches that character and continues the header on the next line. The split character itself is not part of the header.

By default, PROC REPORT uses the slash (/) as the split character.

Note: If you are typing over a header (rather than entering one from the PROMPTER or DEFINITION window), you do not see the effect of the split character until you refresh the screen, either by issuing the REFRESH command or by adding or deleting an item, changing the contents of a DEFINITION window, or changing the contents of a BREAK window.

Panels

creates sets of columns called *panels*. If the width of a report is less than half of the line size, you can display the data in multiple panels so that rows that would otherwise appear on multiple pages appear on the same page. By default, the value of the **Panels** attribute is 1.

The report that appears in Output 9.22 uses two panels.

Output 9.22
Using Panels in a
Report

```
                   Using Two Panels on Each Page of the Report                  1

         QTR  DEPT       ACCOUNT    BUDGET   QTR  DEPT        ACCOUNT    BUDGET
           1  Staff      fulltime  $130,000    1  Facilities  rent      $24,000
           2  Staff      fulltime  $165,000    2  Facilities  rent      $24,000
           1  Staff      parttime   $40,000    1  Facilities  utils      $5,000
           2  Staff      parttime   $60,000    2  Facilities  utils      $3,500
           1  Equipment  lease      $40,000    1  Facilities  supplies   $2,750
           2  Equipment  lease      $40,000    2  Facilities  supplies   $2,750
           1  Equipment  purchase   $40,000    1  Travel      leases     $3,500
           2  Equipment  purchase   $20,000    2  Travel      leases     $4,500
           1  Equipment  tape        $8,000    1  Travel      gas          $800
           2  Equipment  tape       $12,000    2  Travel      gas        $1,200
           1  Equipment  sets        $7,500    1  Other       advert    $30,000
           2  Equipment  sets        $7,500    2  Other       advert    $30,000
           1  Equipment  maint      $10,000    1  Other       talent    $13,500
           2  Equipment  maint      $12,000    2  Other       talent    $19,500
           1  Equipment  rental      $4,000    1  Other       musicfee   $3,000
           2  Equipment  rental      $6,000    2  Other       musicfee   $5,000
```

ROPTIONS *continued*

Pspace

specifies the number of horizontal positions between panels. The default value is 4. The REPORT procedure separates all panels in the report by the same number of horizontal positions. For each panel, the sum of its width and the number of horizontal positions separating it from the panel to its left cannot exceed the line size.

User Help

The fields under **User Help** locate the library and catalog containing the user-defined CBT or help entries for the report. PROC REPORT stores all help entries for a report in the same catalog. You can write a CBT or HELP entry for each item in the report. Specify the entry name from the DEFINITION window for the variable. You create the entry with the BUILD procedure in SAS/AF software. For information on PROC BUILD, see Chapter 10, "The BUILD Procedure," in *SAS/AF Software: Usage and Reference*. The **User Help** fields are described below:

Libname

contains the libref for the SAS data library that contains the help catalog. You can use any valid libref in this field, but be sure to define the libref before invoking the REPORT procedure.

Catalog

contains the name of the catalog that contains the help entries. You can use any valid SAS name in this field.

For more information on SAS data libraries and SAS catalogs, refer to Chapter 6, "SAS Files," in *SAS Language: Reference*.

RSTORE

Prompts you for the complete name of the catalog entry in which to store the definition of the current report

Display

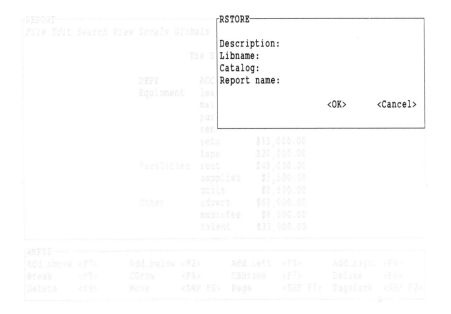

Access

Issue the RSTORE command.

Description

The RSTORE window prompts you for the complete name of the catalog entry in which to store the definition of the current report and for a description of the report. The SAS System stores the report definition as a catalog entry of type REPT.

You can use a report definition to create an identically structured report for any SAS data set that contains variables with the same names as the ones used in the report definition.

For example, you may want to create a report by working interactively with a small data set. When you are satisfied with the report, you can store its definition and use it either interactively or noninteractively to produce a report using a larger data set.

Descriptions of the fields in the RSTORE window follow:

Description

contains a description of the report. This description, which is optional, shows up in the RLOAD window and helps you select the appropriate report. Descriptions can contain up to 40 characters.

RSTORE *continued*

Libname

contains the libref of the SAS data library that contains the catalog in which you want to store the report. You should define a libref with the LIBNAME statement before invoking the REPORT procedure.

Note: Whenever you start a SAS session, the SAS System defines the libref SASUSER for you. Thus, you can always use SASUSER in the **Libname** field.

Catalog

contains the name of the catalog in which you want to store the report. If the catalog does not already exist, the SAS System creates it for you.

Report name

contains the name of the entry to contain the report definition.

STATISTICS

Displays the choices for adding statistics to the report

Display

Access

Select **Statistic** from the ADDING window.

Description

The STATISTICS window displays the statistics you can add to a report. The formulas and definitions for each statistic are given here.

The following notations are used where summation is over all nonmissing values:

x_i the ith nonmissing observation on the variable.

w_i the weight associated with x_i if a WEIGHT statement is specified, and 1 otherwise.

n the number of nonmissing observations.

$\bar{x}$ $\Sigma w_i x_i / \Sigma w_i$.

s^2 $\Sigma w_i (x_i - \bar{x})^2 / (n-1)$.

z_i $(x_i - \bar{x})/s$ standardized variables.

The statistics available are described below:

N   the number of observations with no missing values for a group, order, or across variable; or the number of nonmissing values for an analysis variable.

NMISS   the number of missing values for an analysis variable.

MEAN $\bar{x}$, the arithmetic mean.

STD s, the standard deviation.

MIN   the minimum value.

MAX   the maximum value.

RANGE $\text{MAX} - \text{MIN}$, the range.

SUM $\Sigma w_i x_i$, the weighted sum.

USS $\Sigma w_i x_i^2$, the uncorrected sum of squares.

CSS $\Sigma w_i (x_i - \bar{x})^2$, the sum of squares corrected for the mean.

STDERR $s/\sqrt{n}$, the standard error of the mean.

SUMWGT Σw_i, the sum of weights.

CV $100\, s/\bar{x}$, the percent coefficient of variation.

T $t = \bar{x}^* \sqrt{n}/s$, Student's t for H_0: population mean$=0$.

PRT   the two-tailed p-value for Student's t with $n-1$ degrees of freedom, the probability under the null hypothesis of obtaining an absolute value of t greater than the t-value observed in this sample.

VAR s^2, the variance.

For details on adding statistics to a report, see the adding commands in Chapter 8.

Chapter 10 The REPORT Language

Introduction

Earlier chapters in this book explain how to create reports in a windowing environment using REPORT commands and windows. This chapter describes the REPORT language, which enables you to create a report in a nonwindowing environment or to create a partial report that you can then modify in a windowing environment. When you use the REPORT language, you submit a series of statements with the PROC REPORT statement, just as you do in many other SAS procedures.

Elements of the REPORT language correspond to the commands and choices in windows described in the chapters about using the REPORT procedure in a windowing environment. For example, many of the options available in the PROC REPORT statement are options in the ROPTIONS window. In such cases, this chapter frequently refers you to the documentation for a windowing environment for complete details.

The first part of this chapter presents the high-level syntax for using the REPORT procedure. Following this syntax are brief descriptions of the purpose of each statement that PROC REPORT supports. An individual section on each statement describes the statement in detail, explaining its purpose and how to use it. Finally, the examples show how to use the REPORT language to create the same reports created in a windowing environment in Chapter 2 through Chapter 6.

REPORT Procedure Syntax

The REPORT procedure uses the following statements:

PROC REPORT <*report-option-list*>;
    BREAK BEFORE | AFTER *break-variable* </ *break-option-list*>;
    COLUMN *column-list*;
    DEFINE *item* </ *define-option-list*>;
    RBREAK BEFORE | AFTER </ *break-option-list*>;
    COMPUTE BEFORE | AFTER <*break-variable*>;
    COMPUTE *computed-variable* </ *type-specification*>;
    ENDCOMP;
    LINE *specification-list*;
    BY *variable-list*;
    FREQ *variable*;
    WEIGHT *variable*;

□   The PROC REPORT statement is required.

□   The following statements are optional:

BREAK	BY
COLUMN	FREQ
DEFINE	WEIGHT
RBREAK	

    Note: You can use the BY statement with PROC REPORT only in a nonwindowing environment.

□ The following statements are optional, but if you use one, you must use the other:

□ COMPUTE

□ ENDCOMP

The COMPUTE statement must precede the ENDCOMP statement. You can use many DATA step statements, all functions, and the LINE statement between the COMPUTE and the ENDCOMP statements. For a list of DATA step statements you can use, see Chapter 6, "Using the COMPUTE Window."

□ The LINE statement is optional. You can use it only between a COMPUTE and an ENDCOMP statement.

For complete statement syntax, see the section on the appropriate statement.

Statement Descriptions

This section describes the purpose of each statement that the REPORT procedure supports.

PROC REPORT	invokes the REPORT procedure.
BREAK	controls the REPORT procedure's actions at a break (change) in the value of a group or order variable.
BY	creates a separate report for each BY group. A BY group is a set of observations with the same values for all BY variables.
COLUMN	describes the arrangement of columns and of headers that span more than one column in the report.
COMPUTE	marks the beginning of a group of DATA step statements, LINE statements, or both that define computed variables, perform calculations, or create customized break lines in the report. The ENDCOMP statement ends the group of statements.
DEFINE	defines the characteristics of the specified item. These characteristics include how PROC REPORT uses the item in the report, formats the item, and chooses the column header.
ENDCOMP	ends the group of DATA step statements, LINE statements, or both that follow a COMPUTE statement.
FREQ	specifies a numeric variable in the input data set whose value represents the frequency of each observation.
LINE	writes break lines containing text, values calculated for the report, or both.
RBREAK	controls the REPORT procedure's actions at the beginning or end of the report.
WEIGHT	specifies a numeric variable in the input data set whose value weights each variable in the report.

PROC REPORT Statement

The PROC REPORT statement invokes the REPORT procedure. Optionally, the statement specifies the SAS data set and report definition to use for the report, as well as a number of options that control the layout of the report and the behavior of the REPORT procedure.

Syntax

The general form of the PROC REPORT statement is

PROC REPORT <*report-option-list*>;

where *report-option-list* can be one or more of the following:

CENTER | NOCENTER	OUTREPT=*libref.catalog.entry*
COLWIDTH=*column-width*	PANELS=*number-of-panels*
COMMAND	PROFILE=*libref.catalog*
DATA=*SAS-data-set*	PROMPT
HEADLINE	PS=*page-size*
HEADSKIP	PSPACE=*space-between-panels*
HELP=*libref.catalog*	REPORT=*libref.catalog.entry*
LIST	SHOWALL
LS=*line-size*	SPACING=*space-between-columns*
MISSING	SPLIT='*character*'
NAMED	WINDOWS | NOWINDOWS
NOHEADER	WRAP
NORKEYS	

Options

You can use the following options with the PROC REPORT statement:

CENTER | NOCENTER
> specifies whether or not to center the report in the REPORT window. You can control the centering of a report from the PROC REPORT statement, from a stored report definition, or from the setting of the CENTER system option. The REPORT procedure first honors the setting in the PROC REPORT statement. If you do not specify how to center the report there, PROC REPORT honors the setting of the CENTER option in a report definition loaded with the REPORT= option in the PROC REPORT statement. If you use neither a procedure statement option nor a stored report definition, PROC REPORT honors the setting of the CENTER system option. You can change centering during a PROC REPORT session with the CENTER option in the ROPTIONS window.
>
> **Note:** Loading a report definition during a PROC REPORT session causes the value of the CENTER option stored with the report definition to take effect.

COLWIDTH=*column-width*
> specifies the default number of horizontal positions for all columns in the report. Valid values range from 1 to the line size. The default value is 9.
>
> When setting the width for a column, PROC REPORT first looks for a width specification in the definition for that item. If it finds none and the item is a data set variable, it uses a column width large enough to accommodate the format (if one exists) associated with the variable in the input data set. Otherwise, it uses the value of the COLWIDTH= option.

COMMAND
> displays command lines in all REPORT windows rather than action bars when you invoke the REPORT procedure. After you have invoked the REPORT procedure, you can display the action bars by issuing the COMMAND command.

DATA=*SAS-data-set*
> names the SAS data set to use for input. If you do not specify the DATA= option, PROC REPORT uses the data set whose name is stored in the automatic variable _LAST_. This data set is usually the data set most recently created within your SAS session, although you can set the value of _LAST_ yourself with the _LAST_= system option.

HEADLINE
> underlines all column headers and the spaces between them at the top of each page of the report.

HEADSKIP
> writes a blank line beneath all column headers (or beneath the underlining that the HEADLINE option writes) at the top of each page of the report.

HELP=*libref.catalog*
> names the library and catalog containing the user-defined CBT or HELP entries for the report. You must store all help entries for a report in the same catalog. You can write a CBT or HELP entry for each item in the report. Specify the entry name in the DEFINITION window for the variable or in a DEFINE statement. You create the entry with the BUILD procedure in SAS/AF software. For information on PROC BUILD, see Chapter 10, "The BUILD Procedure," in *SAS/AF Software: Usage and Reference, Version 6, First Edition.*
>
> When you use the HELP= option, you must specify the following:
>
> | | |
> |---|---|
> | *libref* | points to the SAS data library containing the HELP catalog. You can use any valid libref in this field, but be sure to define the libref before invoking the REPORT procedure. Use the LIBNAME statement to associate a libref with the SAS data library. For details on the LIBNAME statement, see *SAS Language: Reference, Version 6, First Edition.* |
> | *catalog* | points to a catalog in *libref.* |

LIST
> writes to the SAS log a listing of the REPORT language that defines the current report. This option has no effect in the windowing environment. For information on a comparable feature in the windowing environment, see the documentation for the LIST command.

LS=*line-size*

specifies the length of a line of the report. Valid values range from 64 to 256. You can control the line size of a report from the PROC REPORT statement, from a stored report definition, or from the LINESIZE= system option. The REPORT procedure first honors the setting in the PROC REPORT statement. If you do not specify a line size for the report there, PROC REPORT honors the setting of the Linesize attribute in a report definition loaded with the REPORT= option in the PROC REPORT statement. If you use neither a procedure statement option nor a stored report definition, PROC REPORT honors the setting of the LINESIZE= system option. You can change the line size during a PROC REPORT session with the Linesize attribute in the ROPTIONS window.

Note: Loading a report definition during a PROC REPORT session causes the value of the Linesize attribute stored with the report definition to take effect.

MISSING

considers missing values as valid values for group, order, or across variables. Special missing values used to represent numeric values (the letters A through Z and the underscore (_) character) are each considered as a different value. A group for each missing value appears in the report. If you do not specify the MISSING option, PROC REPORT does not include observations with a missing value for one or more group, order, or across variables in the report.

NAMED

writes *name=* in front of each value in the report, where *name* is the column header for the value. You may find it useful to specify the NAMED option in conjunction with the WRAP option to produce a report that wraps all columns for a single row of the report onto consecutive lines rather than placing columns of a wide report on separate pages. When you select the NAMED option, PROC REPORT automatically uses the NOHEADER option.

For an example of a report that uses the NAMED option, see the documentation for the NAMED option in the ROPTIONS window.

NOHEADER

suppresses column headers from the report, including those that span multiple columns.

▶ *Caution* . . . . *Once you suppress the display of column headers, you cannot select any report items.* ▲

NORKEYS

suppresses the initial display of the RKEYS window. After you have invoked the REPORT procedure, you can use the RKEYS command to display the RKEYS window.

OUTREPT=*libref.catalog.entry*
>   names the entry in which to store the report definition in use when the REPORT procedure terminates.
>
>   When you use the OUTREPT= option, you must specify the following:
>
>   *libref*   points to a SAS data library. Use the LIBNAME statement to associate a libref with the SAS data library. For details on the LIBNAME statement, see *SAS Language: Reference*.
>
>   *catalog*   names a SAS catalog in *libref*.
>
>   *entry*   names the entry to which to write the report definition.
>
>   The SAS System assigns an entry type of REPT to the entry.
>
>   **Note:** You can use the RSTORE command to save report definitions from the REPORT window before you are ready to terminate the procedure.

PANELS=*number-of-panels*
>   specifies the number of panels on each page of the report. If the width of a report is less than half of the line size, you can display the data in multiple sets of columns so that rows that would otherwise appear on multiple pages appear on the same page. Each set of columns is called a *panel*. A familiar example of this kind of report is a telephone book, which contains multiple panels of names and telephone numbers on a single page.
>
>   By default, the REPORT procedure creates a report with one panel per page. For an example of a report that uses multiple panels, see the Panels attribute in the documentation for the ROPTIONS window.

PROFILE=*libref.catalog*
>   locates your REPORT profile. A profile enables you to customize certain aspects of the REPORT procedure. For information on using a REPORT profile, see the documentation for the PROFILE window.
>
>   When you use the PROFILE= option, you must specify the following:
>
>   *libref*   points to a SAS data library. You can use any valid libref in this field, but be sure to define the libref before invoking the REPORT procedure. Use the LIBNAME statement to associate a libref with the SAS data library. For details on the LIBNAME statement, see *SAS Language: Reference*.
>
>   *catalog*   contains the name of the catalog that contains your profile. You can use any valid SAS name in this field.
>
>   When you use the PROFILE= option, PROC REPORT uses the entry REPORT.PROFILE in the catalog you specify as your profile. If no such entry exists, or if you do not specify a profile, PROC REPORT uses the profile in SASUSER.PROFILE. If you have no profile, PROC REPORT uses the defaults for all options, colors, pull-down menus, and action bars.
>
>   **Note:** When you create a profile, PROC REPORT stores it in SASUSER.PROFILE.REPORT.PROFILE. Use the CATALOG procedure or the CATALOG window if you want to copy the profile to another location.

PROMPT
>opens the REPORT window and starts the PROMPT facility. This facility guides you through creating a new report or adding more data set variables or statistics to an existing report.

PS=*page-size*
>specifies the number of lines in a page of the report. Valid values range from 15 to 32,767. You can control the page size of a report from the PROC REPORT statement, from a stored report definition, or from the setting of the PAGESIZE= system option. The REPORT procedure first honors the setting in the PROC REPORT statement. If you do not specify a page size for the report there, PROC REPORT honors the setting of the Pagesize attribute in a report definition loaded with the REPORT= option in the PROC REPORT statement. If you use neither a procedure statement option nor a stored report definition, PROC REPORT honors the PAGESIZE= system option. You can change the page size during a PROC REPORT session with the Pagesize attribute in the ROPTIONS window.
>
>   **Note:**   Loading a report definition during a PROC REPORT session causes the value of the Pagesize attribute stored with the report definition to take effect.

PSPACE=*space-between-panels*
>specifies the number of blank characters between panels. The default value is 4.
>
>   The REPORT procedure separates all panels in the report by the same number of blank characters. For each panel, the sum of its width and the number of blank characters separating it from the panel to its left cannot exceed the line size.
>
>   For information on altering the amount of space between panels after you have invoked the REPORT procedure in the windowing environment, see the Pspace attribute in the documentation for the ROPTIONS window.

REPORT=*libref.catalog.entry*
>names the report definition to load initially. The REPORT procedure stores all report definitions as entries of type REPT in a SAS catalog. When you use the REPORT= option, you must specify the following:

>| | |
>|---|---|
>| *libref* | points to a SAS data library. Use the LIBNAME statement to associate a libref with the SAS data library. For details on the LIBNAME statement, see *SAS Language: Reference.* |
>| *catalog* | names a SAS catalog in *libref.* |
>| *entry* | names the entry that contains the report definition you want. |

>   If you don't use the REPORT= option and the input data set contains both character and numeric variables, the REPORT procedure creates a report similar to the output generated by a simple PROC PRINT step. If you don't use the REPORT= option and the input data set contains only numeric variables, PROC REPORT produces a one-line summary report that shows the sum of each variable over all observations in the data set. However, in either case, in the windowing environment, PROC REPORT initially uses only as many of the variables in the data set as can fit on one page of the report. You can easily add other variables to the report.

For information on loading a different report after you have invoked the REPORT procedure in a windowing environment, see the documentation for the RLOAD command.

SHOWALL

overrides selections in a report definition (specified by the REPORT= option) that suppress the display of a column. See the NOPRINT and NOZERO options in the documentation for the DEFINE statement.

SPACING=*space-between-columns*

specifies the number of blank characters between columns. The default value is 2.

The REPORT procedure separates all columns in the report by the number of blank characters specified by the SPACING= option in the PROC REPORT statement unless you use the SPACING= option in the DEFINE statement to change the spacing to the left of a specific item. For each column, the sum of its width and the blank characters between it and the column to its left cannot exceed the line size. For information on altering the amount of space used between columns after you have invoked the REPORT procedure in the windowing environment, see the Spacing attribute in the documentation for the ROPTIONS window.

SPLIT=*'character'*

specifies the split character. If you use the split character in a column header, the REPORT procedure breaks the header when it reaches that character and continues the header on the next line. The split character itself is not part of the column header. By default, PROC REPORT uses the slash (/) as the split character.

WINDOWS | NOWINDOWS
WD | NOWD

selects a windowing or nonwindowing environment. The default is NOWINDOWS. When you use the WINDOWS option, the SAS System opens the REPORT window, which enables you to modify a report repeatedly and to see the modifications immediately. When you use the NOWINDOWS option, PROC REPORT runs without the REPORT window and sends its output to the SAS procedure output.

WRAP

displays one value from each column of the report, on consecutive lines if necessary, before displaying another value from the first column. By default, PROC REPORT displays values only for as many columns as it can fit on one page. It fills a page with values for these columns before starting to display values for the remaining columns on the next page.

For an example of a report that uses the WRAP and NAMED options, see the NAMED option in the documentation for the ROPTIONS window.

BREAK Statement

The BREAK statement controls the REPORT procedure's actions at a break (change) in the value of a group or order variable. For a description of breaks and break lines, see the documentation for the BREAK window and the BREAK command.

Syntax

The general form of the BREAK statement is

BREAK BEFORE | AFTER *break-variable* </ *break-option-list*>;

where

□ *break-variable* is a group or order variable. The REPORT procedure performs the actions you specify each time the value of this variable changes.

□ *break-option-list* can be one or more of the following:

COLOR=*color*	SKIP
DOL	SUMMARIZE
DUL	SUPPRESS
OL	UL
PAGE	

Requirements

A BREAK statement must include a location, BEFORE or AFTER, that specifies the location of the break lines you are defining, and a break variable. The location must precede the break variable.

BEFORE
  places the break lines before the first row of each set of rows that have the same value for the break variable.

AFTER
  places the break lines after the last row of each set of rows that have the same value for the break variable.

For illustrations of placing break lines before and after each set of rows, see the documentation for the BREAK command.

Options

You can use the following options in the BREAK statement:

COLOR=*color*
 specifies the color of the break lines. Valid values for *color* are

BLACK	GRAY	PINK
BLUE	GREEN	RED
BROWN	MAGENTA	WHITE
CYAN	ORANGE	YELLOW

 The default color is the color of the text in the report.
 Note: Not all operating systems and devices support all colors, and on some operating systems and devices, one color may map to another color. For example, if the BREAK window displays BROWN in yellow characters, selecting BROWN results in yellow break lines.
 Currently, color appears only when PROC REPORT displays the report in the REPORT window.

DOL
 (for double overlining) writes a line of equals signs (=) across the report above each value that appears in the summary line or that would appear in the summary line if you specified the SUMMARIZE option. If you specify both OL and DOL, PROC REPORT honors only the OL option.

DUL
 (for double underlining) writes a line of equals signs (=) across the report below each value that appears in the summary line or that would appear in the summary line if you specified the SUMMARIZE option. If you use both the DUL and the UL options, PROC REPORT honors only the UL option.

OL
 (for overlining) writes a line of hyphens (—) across the report above each value that appears in the summary line or that would appear in the summary line if you specified the SUMMARIZE option.

PAGE
 starts a new page after the last line of each group of break lines.

SKIP
 writes a blank line after the last line of each group of break lines.

SUMMARIZE
 writes a summary line (a line containing summaries of statistics and computed variables) in each group of break lines.

SUPPRESS

suppresses printing of both the value of the break variable in the summary line and any underlining and overlining in the break lines in the column containing the break variable.

Note: If you use the SUPPRESS option, the value of the break variable is unavailable for use in customized break lines unless you assign it a value in the COMPUTE window associated with the break.

UL

(for underlining) writes a line of hyphens (—) across the report below each value that appears in the summary line or that would appear in the summary line if you specified the SUMMARIZE option.

When a break contains more than one break line, the order in which the break lines appear is as follows:

1. overlining or double overlining (OL or DOL)

2. summary line (SUMMARIZE)

3. underlining or double underlining (UL or DUL)

4. break lines defined in a code segment attached to the break through the COMPUTE and ENDCOMP statements

5. skipped line (SKIP).

BY Statement

The BY statement creates a separate report for each BY group. A BY group is a set of observations with the same values for all BY variables. The general form of the BY statement is

BY *variable-list*;

When you use a BY statement, the REPORT procedure expects the input data set to be sorted in order of the BY variables or to have an appropriate index. If your input data set is not sorted in ascending order, you can do one of the following:

□ Use the SORT procedure with a similar BY statement to sort the data.

□ If appropriate, use one of the BY statement options NOTSORTED or DESCENDING.

□ Create an index on the BY variables you want to use. For more information on creating indexes and using the BY statement with indexed data sets, see Chapter 17, "The DATASETS Procedure," in the *SAS Procedures Guide, Version 6, Third Edition.*

▶ *Caution . . . .* **You cannot use the BY statement in the windowing environment.** ▲

COLUMN Statement

The COLUMN statement describes the arrangement of columns and of headers that span more than one column in the report.

Syntax

The general form of the COLUMN statement is

COLUMN *column-list*;

where *column-list* is a concatenation of one or more lists of items that define the columns of the report. Each list of items may consist of the following:

□ the name of a single item: a data set variable, a computed variable, or a statistic. Valid statistics are

N	RANGE	T
NMISS	SUM	PRT
MEAN	USS	SUMWGT
STD	CSS	VAR
MIN	STDERR	
MAX	CV	

For example, the following COLUMN statement creates a report with three columns: one for DEPT, one for BUDGET, and one for ACTUAL:

```
column dept budget actual;
```

□ two or more items separated by commas. The REPORT procedure stacks items separated by commas one above another in the report, with the leftmost item on top. For example, the following COLUMN statement creates a report with three columns: one for department, one with the MEAN statistic below BUDGET, and one with the MEAN statistic below ACTUAL. The DEFINE statement defines DEPT as a group variable so that the REPORT procedure can calculate the MEAN statistics. The results appear in Output 10.1.

```
libname report 'SAS-data-library';
proc report data=report.budget;
   column dept budget,mean actual,mean;
   define dept / group;
run;
```

Output 10.1
Stacking Items
with the COLUMN
Statement

```
                       The SAS System                          1

                         BUDGET        ACTUAL
           DEPT           MEAN          MEAN
           Equipment     $17,250       $17,369
           Facilities    $10,333       $10,105
           Other         $16,833       $18,107
           Staff         $98,750       $98,464
           Travel         $2,500        $2,114
```

□ two or more lists of items in parentheses. The REPORT procedure applies any item separated from a parenthesized list by a comma to each list inside the parentheses. For example, the following COLUMN statements are equivalent:

```
column dept (budget actual),mean;
```

```
column dept budget,mean actual,mean;
```

You can also use parentheses to specify headers that span multiple columns. In such a case, one or more of the items in a list is a quoted string to use as the header:

(*'header-1'* < . . . *'header-n'*> *item-list*)

where *header* is a string of characters that spans one or more columns in the report. The list of items following the headers within the same pair of parentheses specifies the columns to span. The REPORT procedure prints each header on a separate line.

You can use split characters in a header to split one header over multiple lines. See the Split attribute in the documentation for the ROPTIONS window for information on using the split character.

If the first and last characters of a header are —, =, _, ., *, or +, or if the first and last characters of a header are < and > or > and <, the REPORT procedure expands the header by repeating the first and last characters to fill the space over the columns the header spans.

For example, the following statements enhance the report in Output 10.1 by adding a two-line header that includes the date of the report (accessed by the automatic macro variable SYSDATE) above the columns for BUDGET and ACTUAL. The enhanced report is in Output 10.2.

```
libname report 'SAS-data-library';
```

```
proc report data=report.budget;
    column dept ("Report as of" "_&sysdate._" budget,mean actual,mean);
    define dept / group;
run;
```

Output 10.2
Creating a Header
That Spans
Multiple Columns

```
                                 The SAS System                              1

                                    Report as of
                                   ___15AUG90___
                                   BUDGET        ACTUAL
                    DEPT            MEAN          MEAN
                    Equipment       $17,250       $17,369
                    Facilities      $10,333       $10,105
                    Other           $16,833       $18,107
                    Staff           $98,750       $98,464
                    Travel           $2,500        $2,114
```

Note: When you use a macro variable in a header, be sure to enclose the header in double rather than single quotes so that the SAS System can resolve the macro variable. Furthermore, if you specify one of the special characters that expands the header, place a period after the name of the macro variable so that the special character does not become part of the name of the macro variable. For more information on using the SAS macro facility, see the *SAS Guide to Macro Processing, Version 6, Second Edition*.

You can use the same data set variable, computed variable, or statistic more than once in the same COLUMN statement. However, you can use only one DEFINE statement for any given name. Therefore, you may need to create an alias for a variable or statistic that you use multiple times in a report. You can then use one DEFINE statement for each alias. (The DEFINE statement defines characteristics such as the usage of the item in the report and what its format is. If you do not use a DEFINE statement for an item, the REPORT procedure uses default values for the characteristics.)

The COLUMN statement used to create the report in Output 10.2 includes two uses of the MEAN statistic, one attached to the variable BUDGET and one attached to the variable ACTUAL:

```
column dept ("Report as of" "_&sysdate._" budget,mean actual,mean);
```

In both cases, the default characteristics are acceptable, so you don't need DEFINE statements for the MEAN statistics.

But suppose that you want to assign different formats and headers to each occurrence of the MEAN statistic. You specify headers and formats with the DEFINE statement. Because you can only have one DEFINE statement for the name MEAN, you need to create an alias for at least one occurrence of MEAN so that you can use separate DEFINE statements to specify the formats and headers.

You assign an alias in the COLUMN statement by replacing the name of an item with the following:

name=alias

where *name* is the name of the item, and *alias* is the alias to use in the DEFINE statement. For example, the following statement provides aliases for both occurrences of the MEAN statistic:

```
column dept budget,mean=budmean actual,mean=actmean;
```

Now, you can use BUDMEAN and ACTMEAN in separate DEFINE statements to assign them different characteristics. For example, the following SAS statements create a report that uses different formats and headers for the two occurrences of the MEAN statistic. The results appear in Output 10.3.

```
libname report 'SAS-data-library';

proc report data=report.budget;

   column dept budget,mean=budmean actual,mean=actmean;

   define dept / group;
   define budmean / format=dollar8. 'Average/Budgeted';
   define actmean / format=dollar11.2 'Average/Spent';
run;
```

Output 10.3
Using Aliases in
the COLUMN
Statement

```
                                   The SAS System                              1

                                    BUDGET         ACTUAL
                                   Average        Average
                      DEPT         Budgeted         Spent
                      Equipment    $17,250      $17,369.10
                      Facilities   $10,333      $10,104.52
                      Other        $16,833      $18,106.74
                      Staff        $98,750      $98,464.38
                      Travel        $2,500       $2,114.25
```

Note: Assigning an alias in the COLUMN statement does not by itself alter the report. It simply enables you to use separate DEFINE statements for each occurrence of a variable or statistic.

COMPUTE Statement

The COMPUTE statement marks the beginning of a group of DATA step statements, DATA step functions, LINE statements, or a combination of these that define computed variables, perform calculations, or create customized break lines. An ENDCOMP statement marks the end of the group of statements. This group of statements is a *code segment*.

Syntax

The general form of the COMPUTE statement is

COMPUTE BEFORE | AFTER <*break-variable*>;
COMPUTE *computed-variable* </ *type-specification*>;

where

□ *break-variable* is a group or order variable.

□ *computed-variable* is a computed variable in the report. You must include the variable in the COLUMN statement (see "Executing Code Segments" later in this section), and you must include a DEFINE statement for each computed variable.

□ *type-specification*, which indicates that the computed variable is a character variable, is either or both of the following:

CHARACTER | CHAR
LENGTH=*length*

Requirements

An ENDCOMP statement must follow the last DATA step statement or LINE statement.

If you specify a location (BEFORE or AFTER) in the COMPUTE statement, you associate DATA step statements, LINE statements, or both with a break at a particular part of the report. The location determines when PROC REPORT executes the statements between the COMPUTE and ENDCOMP statements:

BEFORE

> executes the statements between the COMPUTE and ENDCOMP statements at a break in one of the following places:

> □ before the first row of a set of rows that have the same value for *break-variable*.

> □ at the beginning of the report if you do not specify a break variable.

AFTER

> executes the statements between the COMPUTE and ENDCOMP statements at a break in one of the following places:

> □ after the last row of a set of rows that have the same value for *break-variable*.

> □ at the end of the report if you do not specify a break variable.

> For examples of breaks before and after a break variable, see the documentation for the BREAK command. For examples of breaks at the beginning and end of a report, see the documentation for the RBREAK command.

> If you specify a computed variable in the COMPUTE statement, you associate the DATA step statements between the COMPUTE and the ENDCOMP statements with a computed variable in the report. You must include the computed variable in the COLUMN statement (see "Executing Code Segments" later in this section), and you must include a DEFINE statement for each computed variable.

> **Note:** When you use the COMPUTE statement, you don't necessarily have to use a corresponding BREAK or RBREAK statement. You need these statements only when you want to implement one or more BREAK statement or RBREAK statement options.

Options

If you specify a location (BEFORE or AFTER) in the COMPUTE statement, you can also specify a *break-variable*. When you specify a break variable, the REPORT procedure executes the statements between the COMPUTE and ENDCOMP statements each time the value of the break variable changes.

If you specify a computed variable in the COMPUTE statement, you can also specify one or both of the following options. If you specify neither option, PROC REPORT assumes that the computed variable is numeric.

CHARACTER | CHAR

> specifies that the computed variable is a character variable. The default is numeric.

LENGTH=*length*

> specifies the length of the character variable. Valid values for *length* range from 1 to 200. If you specify CHARACTER but do not specify a length, the default length is 8.

Executing Code Segments

In order to use the COMPUTE statement successfully, you must understand a few details about when PROC REPORT executes the statements between a COMPUTE statement and its corresponding ENDCOMP statement. Chapter 6 illustrates this information in more detail.

Placement of Computed Variables in the COLUMN Statement

When you use the COMPUTE statement to attach DATA step statements to a computed variable, the REPORT procedure executes these statements once for each row of the report. The value of a computed variable that appears in a report for any given row is the last value assigned to that variable's name during that execution of the DATA step statements. Because the REPORT procedure writes the rows of a report from left to right, a computed variable can depend only on values of items that appear to its left in the report. You must place the variable accordingly in the COLUMN statement.

Placement of Break Lines

When you use a location to attach DATA step statements and LINE statements to a break, the REPORT procedure executes these statements once at each break. Because PROC REPORT calculates statistics for groups before it actually constructs the rows of a report, statistics for sets of detail rows are available before or after the detail rows are displayed, as are values for any variables based on these statistics. For instance, if you want to display the mean of the variable BUDGET, add the MEAN statistic above or below BUDGET, or use the DEFINE statement or the DEFINITION window to associate the mean statistic with BUDGET. The mean is then available in the variable BUDGET.MEAN at the break, whether the break is before or after the detail rows. You can store this value in another variable whose value won't change as PROC REPORT builds the report (the value of BUDGET.MEAN changes in each row) and use it to calculate other values in the report. This technique is particularly useful for calculating percentages, such as the percentage of the budget allocated to each account in a department. For an example of calculating percentages, see "Adding a Customized Break Line" in Chapter 5, "Using Some Advanced Features of the REPORT Procedure."

While summary statistics are available independent of the location of the break, some values are not available until the REPORT procedure has constructed the last detail row of a group. For example, you may want to determine the maximum difference between BUDGET and ACTUAL within each department. In such a case, you must compute this difference for each detail row and determine in which row the difference is greatest. The REPORT procedure cannot know which difference is greatest until it has processed the entire set of rows. Therefore, if you want to use the maximum difference in any break lines, the break lines must appear after the detail rows.

Note: You can initialize variables by breaking before the detail rows over which you want to use the variable. For an example of this technique see "Calculating Values for a Set of Rows" in Chapter 6.

Selection of DATA Step Statements

You can use a wide variety of DATA step statements between the COMPUTE and ENDCOMP statements. The REPORT procedure provides an additional statement, the LINE statement, that writes text, including the values of items in the report, in break lines. See the documentation for the LINE statement for details.

Chapter 6 provides details on the use of DATA step statements in the COMPUTE window.

DEFINE Statement

The DEFINE statement defines the characteristics of the specified item. These characteristics include how PROC REPORT uses the item in the report, formats the item, and chooses the column header.

Syntax

The general form of the DEFINE statement is

DEFINE *item*
        $</$ *<usage>*
        <attribute-list>
        <option-list>
        <justification>
        <'column-header-1' <. . . 'column-header-n'>>
        *<*COLOR=*color>>*;

where

- □ *item* is the name or alias (established in the COLUMN statement) of a data set variable, a computed variable, or a statistic

- □ *usage* can be one of the following:

ACROSS	DISPLAY
ANALYSIS	GROUP
COMPUTED	ORDER

- □ *attribute-list* can be one or more of the following:

FORMAT=*format*	*statistic*
ITEMHELP=*entry-name*	WIDTH=*column-width*
SPACING=*horizontal-positions*	

- □ *option-list* can be one or more of the following:

  DESCENDING
  NOPRINT
  NOZERO
  PAGE

- □ *justification* can be one of the following:

  CENTER
  LEFT
  RIGHT

Requirements

A DEFINE statement must include the argument *item*, which is the item in the report whose characteristics the DEFINE statement describes. An item can be a data set variable, a computed variable, a statistic, or an alias from a COLUMN statement.

Options

You can use the following options in the DEFINE statement:

ACROSS
> defines the specified item as an across variable. See the documentation for the DEFINITION window for an explanation of different types of usage.

ANALYSIS
> defines the specified item as an analysis variable. See the documentation for the DEFINITION window for an explanation of different types of usage. If you use the ANALYSIS option, you must specify a statistic (see the *statistic* option). Because the *statistic* option implies the ANALYSIS option, you never need to specify ANALYSIS, although specifying it may make your code easier for novice users to read.
>
> By default, PROC REPORT uses numeric variables as analysis variables used to calculate the SUM statistic.

CENTER
> centers the formatted values of the specified item within the column width and centers the column header over the values. Selecting this option does not alter the setting of the CENTER system option.

COLOR=*color*
> specifies the color to use to display the column header and the values of the specified item. Valid values for *color* are

BLACK	GRAY	PINK
BLUE	GREEN	RED
BROWN	MAGENTA	WHITE
CYAN	ORANGE	YELLOW

> The default color is the color of the text in the report.
>
> **Note:** Not all operating systems and devices support all colors, and on some operating systems and devices, one color may map to another color. For example, if the DEFINITION window displays BROWN in yellow characters, selecting BROWN results in yellow values and a yellow column header.
>
> Currently, color appears only when PROC REPORT displays the report in the REPORT window.

column-header
> defines the column header for the specified item. You must enclose each header in single or double quotation marks. When you specify multiple column headers, PROC REPORT uses a separate line for each one. You can also use the split character to split a column header over multiple lines. See the Split attribute in the documentation for the ROPTIONS window for information on using the split character.

By default, PROC REPORT uses a variable's label as its column header and the name of the statistic as the column header for a statistic. If a variable has no label, PROC REPORT uses the variable's name. If you want to use names when labels exist, submit the following SAS statement before invoking the REPORT procedure:

```
options nolabel;
```

If the first and last characters of a header are —, =, _, ., *, or +, or if the first and last characters of a header are < and > or > and <, the REPORT procedure expands the header by repeating the first and last characters to fill the space over the column.

COMPUTED
   defines the specified item as a computed variable. See the documentation for the DEFINITION window for an explanation of different types of usage.

DESCENDING
   reverses the order in which PROC REPORT displays rows or values of a group, order, or across variable. For instance, when DEPT is the leftmost order variable, the REPORT procedure displays the rows in ascending (alphabetic) order of the value of DEPT. When the characteristics for DEPT include DESCENDING, the values appear in reverse alphabetic order.
   Note: PROC REPORT orders group, order, and across variables by their formatted values.

DISPLAY
   defines the specified item as a display variable. See the documentation for the DEFINITION window for an explanation of different types of usage. By default, PROC REPORT uses character variables as display variables.

FORMAT=*format*
   assigns a SAS or user-defined format to the item. This format applies to the specified item as PROC REPORT displays it; the format does not alter the format stored with a variable in the data set. By default, PROC FORMAT uses the format you specify in the FORMAT statement when you invoke the REPORT procedure. If you do not specify a format, PROC REPORT uses the format stored in the data set. If no format for the item is stored in the data set, it uses the BESTw. format for numeric variables and the $w. format for character variables. The REPORT procedure uses the default column width (specified by the COLWIDTH= option) as the default.

GROUP
   defines the specified item as a group variable. See the documentation for the DEFINITION window for an explanation of different types of usage.

ITEMHELP=*entry-name*
   references a HELP or CBT entry for help on the selected item. Use the BUILD procedure to create a HELP or CBT entry for a report item. (For information on PROC BUILD, see Chapter 10 in *SAS/AF Software: Usage and Reference*.) All HELP and CBT entries for a report must be in the same catalog, and you must specify that catalog with the HELP= option when you invoke the REPORT procedure.

(ITEMHELP= continued)

Of course, you can access these entries only from a windowing environment. To access a help entry from the report, select the item and issue the HELP command. The REPORT procedure first searches for and displays an entry named *entry-name*.CBT. If no such entry exists, it searches for *entry-name*.HELP.

LEFT

left-justifies the formatted values of the specified item within the column width and left-justifies the column headers over the values. If the format width is the same as the width of the column, the LEFT option has no effect.

NOPRINT

suppresses the display of the selected column. Use this option if you do not want to show the column in the report but you need to use the values in it to calculate other values you use in the report or to establish the order of rows in the report.

The SHOWALL option in the PROC REPORT statement overrides all occurrences of the NOPRINT option.

NOZERO

suppresses the display of a column containing values of an analysis or computed variable if all values are zero or missing.

The SHOWALL option in the PROC REPORT statement overrides all occurrences of the NOZERO option.

ORDER

defines the specified item as an order variable. See the documentation for the DEFINITION window for an explanation of different types of usage.

PAGE

inserts a page break in the report just before the first column containing values of the selected item. For an illustration of the PAGE option, see the documentation for the DEFINITION window.

RIGHT

right-justifies the formatted values of the specified item within the column width and right-justifies the column headers over the values. If the format width is the same as the width of the column, RIGHT has no effect.

SPACING=*horizontal-positions*

defines the number of blank characters to leave between the selected column and the column immediately to its left. The default value is 2.

statistic

associates a statistic with an analysis variable. You cannot use a statistic name with any other kind of variable, and you must use a statistic name with an analysis variable (see analysis variables in the documentation for the DEFINITION window).

Note: The REPORT procedure uses the name of the analysis variable as the default header for the column. You can add the name of the statistic to the column header by including the statistic name in the appropriate place in the DEFINE statement or by modifying the column header.

The following are valid values for *statistic*:

N	RANGE	CV
NMISS	SUM	T
MEAN	SUMWGT	PRT
STD	USS	VAR
MIN	CSS	
MAX	STDERR	

For definitions of these statistics, see the documentation for the STATISTICS window.

WIDTH=*column-width*

defines the width of the column in which PROC REPORT displays the selected item. The value for *column-width* can range from 1 to the value of the LINESIZE= system option. By default, the REPORT procedure assigns a column width just large enough to handle the specified format.

Note: When you stack items in the same column in a report, the width of the item that is closest to the first row of the report determines the width of the column. For an illustration of a report that stacks items with different widths, see the documentation for the DEFINITION window.

ENDCOMP Statement

The ENDCOMP statement ends the series of DATA step statements, LINE statements, or both that follow a COMPUTE statement.

The general form of the ENDCOMP statement is

ENDCOMP;

A COMPUTE statement must precede the ENDCOMP statement.

FREQ Statement

The FREQ statement specifies a numeric variable in the input SAS data set whose value represents the frequency of each observation. The general form of the FREQ statement is

FREQ *variable*;

If you use the FREQ statement, PROC REPORT assumes that each observation in the input data set represents *n* observations, where *n* is the value of the FREQ variable. If *n* is not an integer, the REPORT procedure truncates it. If *n* is less than 1 (which includes missing), PROC REPORT skips the observation.

You can use the FREQ statement in combination with a WEIGHT statement.

The information from the FREQ statement is not stored in a report definition.

LINE Statement

The LINE statement, which can only appear between the COMPUTE and ENDCOMP statements, writes break lines containing text, values calculated for a set of rows of the report, or both. The LINE statement provides a subset of the features of the PUT statement, which you use in the DATA step to write lines to the SAS log, the SAS output file, or an external file.

Syntax

The general form of the LINE statement is

LINE *specification-list*;

where *specification-list* can be one or more of the following:

item item-format

'*character-string*'

number-of-repetitions'*character-string*'

pointer-control

Requirements

You must include a *specification-list*. You can use the following kinds of specifications:

item item-format
    specifies the item to display and the format to use to display it, where

item	is the name of a data set variable, a computed variable, or a statistic in the report. When you specify a statistic calculated on the values of an analysis variable, use the compound name that identifies both the statistic and the variable with which it is associated. For example, to write the SUM statistic for BUDGET, use the name BUDGET.SUM.
item-format	is a SAS or user-defined format. You must specify a format for each item.

'*character-string*'
    specifies a string of text to display. When the string is a blank and nothing else is in the *specification-list*, PROC REPORT prints a blank line.

number-of-repetitions'*character-string*'
    specifies a character string and the number of times to repeat it. For example, the following LINE statement displays the text 'End of Report   ' four times:

```
line 4*'End of Report      ';
```

pointer-control

specifies the column in which the REPORT procedure displays the next specification in the list. You can use either of the following forms for pointer controls:

@*column-number* specifies the number of the column in which to begin displaying the next item in the specification list.

+*column-increment* specifies the number of columns to skip before beginning to display the next item in the specification list.

For example, the following LINE statement displays BUDGET.SUM with a DOLLAR11.2 format starting in column 25, then skips 10 columns and displays ACTUAL.SUM with the same format:

```
line @25 budget.sum dollar11.2 +10 actual.sum dollar11.2;
```

Note: The LINE statement does not support the following features of the PUT statement:

□ automatic labeling signaled by an equals sign (=), also known as named output

□ the _ALL_, INFILE, and PAGE arguments and the OVERPRINT option

□ grouping items and formats to apply one format to a list of items

□ pointer control using variables and expressions

□ line pointer controls (# and /)

□ trailing at signs (@ and @@)

□ format modifiers

□ array elements.

RBREAK Statement

The RBREAK statement controls the REPORT procedure's actions at the beginning or end of the report. For a description of breaks and break lines, see the documentation for the BREAK window and the BREAK command.

Syntax

The general form of the RBREAK statement is

RBREAK BEFORE | AFTER </ *break-option-list*>;

where *break-option-list* can be one or more of the following:

COLOR=*color*	OL	SUMMARIZE
DOL	PAGE	UL
DUL	SKIP	

Requirements

An RBREAK statement must include an argument, BEFORE or AFTER, that specifies the location of the break lines you are defining.

BEFORE
  places the break lines at the beginning of the report.

AFTER
  places the break lines at the end of the report.

Options

You can use the following options in the RBREAK statement.

COLOR=*color*
  specifies the color to use to display the column header and the values of the specified item. Valid values for *color* are

BLACK	GRAY	PINK
BLUE	GREEN	RED
BROWN	MAGENTA	WHITE
CYAN	ORANGE	YELLOW

  The default color is white.
    Note: Not all operating systems and devices support all colors, and on some operating systems and devices, one color may map to another color. For example, if the BREAK window displays BROWN in yellow characters, selecting BROWN results in yellow break lines.
    Currently, color appears only when PROC REPORT displays the report in the REPORT window.

DOL
  (for double overlining) writes a line of equals signs (=) across the report above each value that appears in the summary line or that would appear in the summary line if you specified the SUMMARIZE option. If you use both the DOL and the OL options, PROC REPORT honors only the OL option.

DUL
  (for double underlining) writes a line of equals signs (=) across the report below each value that appears in the summary line or that would appear in the summary line if you specified the SUMMARIZE option. If you use both the DUL and the UL options, PROC REPORT honors only the UL option.

OL
  (for overlining) writes a line of hyphens (—) across the report above each value that appears in the summary line or that would appear in the summary line if you specified the SUMMARIZE option.

PAGE
  starts a new page after the last break line of a break located at the beginning of the report.

SKIP

writes a blank line after the last break line of a break located at the beginning of the report.

SUMMARIZE

includes a summary line.

UL

(for underlining) writes a line of hyphens (—) across the report below each value that appears in the summary line or that would appear in the summary line if you specified the SUMMARIZE option.

The order in which break lines appear in a report is

1. overlining or double overlining (OL or DOL)

2. summary line (SUMMARIZE)

3. underlining or double underlining (UL or DUL)

4. break lines defined in a code segment attached to the break through the COMPUTE and ENDCOMP statements

5. skipped line (SKIP).

WEIGHT Statement

The WEIGHT statement specifies a numeric variable in the input data set whose value is used to weight each analysis variable. The general form of the WEIGHT statement is

WEIGHT *weight-variable*;

The WEIGHT variable need not be an integer and does not affect the degrees of freedom. If the value of *weight-variable* is less than 0 (which includes missing), PROC REPORT uses a value of 0.

If you use a WEIGHT statement, PROC REPORT uses the value of the WEIGHT variable to calculate weighted statistics. For information on how the WEIGHT value affects the calculation of statistics, refer to w_i in the documentation for the STATISTICS window.

The information from the WEIGHT statement is not stored in a report definition.

Examples

The examples in this section show the REPORT language needed to create the reports in the tutorials in Chapter 2 through Chapter 6 of this book. The numbers at the ends of the lines of code correspond to the numbers in the numbered list that follows the code.

Example 1: Ordering the Rows of a Report

This example creates a simple report in which PROC REPORT orders the rows first by their values of QTR, then by their values of DEPT, and finally, by their values of ACCOUNT. The report appears in Output 10.4.

The REPORT language that creates this report follows:

```
libname report 'SAS-data-library';

proc report data=report.budget;          1

   column  qtr dept account budget actual;          2

   define  qtr      / order format=1. width=7 'Quarter';          3
   define  dept     / order format=$10. width=10 'Department';
   define  account  / order format=$8. width=8 'Account';

   define  budget   / display format=dollar11.2 width=11          4
                       'Amount/Budgeted';
   define  actual   / display format=dollar11.2 width=11
                       'Amount/Spent';

run;
```

1. Invoke the REPORT procedure.

2. Create one column for each variable in the data set.

3. Define QTR, DEPT, and ACCOUNT as order variables. Assign appropriate formats, widths, and column headers.

4. Define BUDGET and ACTUAL as display variables. Assign appropriate formats, widths, and two-line column headers.

Output 10.4
Ordering the Rows of a Report

```
                              The SAS System                              1

                                        Amount        Amount
         Quarter Department Account     Budgeted       Spent
             1   Equipment  lease      $40,000.00    $40,000.00
                            maint      $10,000.00     $7,542.13
                            purchase   $40,000.00    $48,282.38
                            rental      $4,000.00     $3,998.87
                            sets        $7,500.00     $8,342.68
                            tape        $8,000.00     $6,829.42
                 Facilities rent       $24,000.00    $24,000.00
                            supplies    $2,750.00     $2,216.55
                            utils       $5,000.00     $4,223.29
                 Other      advert     $30,000.00    $32,476.98
                            musicfee    $3,000.00     $2,550.50
                            talent     $13,500.00    $12,986.73
                 Staff      fulltime  $130,000.00   $127,642.68
                            parttime   $40,000.00    $43,850.12
                 Travel     gas           $800.00       $537.26
                            leases      $3,500.00     $3,045.15
```

```
        2  Equipment   lease     $40,000.00   $40,000.00
                       maint     $12,000.00   $10,675.29
                       purchase  $20,000.00   $17,769.15
                       rental     $6,000.00    $5,482.94
                       sets       $7,500.00    $8,079.62
                       tape      $12,000.00   $11,426.73
           Facilities  rent      $24,000.00   $24,000.00
                       supplies   $2,750.00    $2,742.48
                       utils      $3,500.00    $3,444.81
           Other       advert    $30,000.00   $37,325.64
                       musicfee   $5,000.00    $4,875.95
                       talent    $19,500.00   $18,424.64
           Staff       fulltime $165,000.00  $166,345.75
                       parttime  $60,000.00   $56,018.96
           Travel      gas        $1,200.00     $984.93
                       leases     $4,500.00    $3,889.65
```

Example 2:   Enhancing the Ordered Report

This example creates an enhanced version of the report in the previous example. The enhancements are

□   a customized title

□   a visual break, consisting of underlining and a blank line, between the column headers and the first row of the report

□   a new variable, BALANCE, calculated from variables in the data set

□   a page break between quarters.

The report appears in Output 10.5. The REPORT language that creates this report follows:

```
libname report 'SAS-data-library';

proc report data=report.budget headline headskip;      1

   title 'Report of Departments';        2
   title2 'by Quarter';

   column qtr dept account budget actual balance;      3

   define   qtr     / order format=1. width=7 'Quarter';      4
   define   dept    / order format=$10. width=10 'Department';
   define   account / order format=$8. width=8 'Account';
   define   budget  / display format=dollar11.2 width=11
                      'Amount/Budgeted';
   define   actual  / display format=dollar11.2 width=11
                      'Amount/Spent';

   define   balance / computed format=dollar11.2 width=11      5
                      'Balance';
```

```
compute balance;      6
   balance=budget-actual;
endcomp;

break after qtr / page;      7

run;
```

1. Invoke the report procedure. The HEADLINE and HEADSKIP options print a row of underlining and a blank line between the column headers and the first detail row.

2. Customize the title of the report.

3. Create one column for each variable in the data set and one column for the computed variable BALANCE.

4. Define the data set variables as in the previous example.

5. Define a computed variable named BALANCE with a format of DOLLAR11.2, a width of 11, and a column header of 'Balance'.

6. Assign a value to the computed variable, BALANCE.

7. Put a page break in the report each time the value of QTR changes. The page break occurs after the last row in each set of rows that have the same value of QTR.

Output 10.5
Enhancing the
Ordered Report

```
                               Report of Departments                        1
                                   by Quarter

                                         Amount      Amount
         Quarter Department  Account    Budgeted      Spent      Balance
         -----------------------------------------------------------------

            1   Equipment   lease      $40,000.00  $40,000.00       $0.00
                            maint      $10,000.00   $7,542.13   $2,457.87
                            purchase   $40,000.00  $48,282.38  $-8,282.38
                            rental      $4,000.00   $3,998.87       $1.13
                            sets        $7,500.00   $8,342.68    $-842.68
                            tape        $8,000.00   $6,829.42   $1,170.58
                Facilities  rent       $24,000.00  $24,000.00       $0.00
                            supplies    $2,750.00   $2,216.55     $533.45
                            utils       $5,000.00   $4,223.29     $776.71
                Other       advert     $30,000.00  $32,476.98  $-2,476.98
                            musicfee    $3,000.00   $2,550.50     $449.50
                            talent     $13,500.00  $12,986.73     $513.27
                Staff       fulltime  $130,000.00 $127,642.68   $2,357.32
                            parttime   $40,000.00  $43,850.12  $-3,850.12
                Travel      gas           $800.00     $537.26     $262.74
                            leases      $3,500.00   $3,045.15     $454.85
```

```
                         Report of Departments                          2
                             by Quarter

                              Amount       Amount
    Quarter Department  Account  Budgeted       Spent      Balance
    -----------------------------------------------------------------

       2  Equipment   lease    $40,000.00   $40,000.00       $0.00
                      maint    $12,000.00   $10,675.29   $1,324.71
                      purchase $20,000.00   $17,769.15   $2,230.85
                      rental    $6,000.00    $5,482.94     $517.06
                      sets      $7,500.00    $8,079.62    $-579.62
                      tape     $12,000.00   $11,426.73     $573.27
          Facilities  rent     $24,000.00   $24,000.00       $0.00
                      supplies  $2,750.00    $2,742.48       $7.52
                      utils     $3,500.00    $3,444.81      $55.19
          Other       advert   $30,000.00   $37,325.64   $-7,325.64
                      musicfee  $5,000.00    $4,875.95     $124.05
                      talent   $19,500.00   $18,424.64   $1,075.36
          Staff       fulltime $165,000.00 $166,345.75  $-1,345.75
                      parttime $60,000.00   $56,018.96   $3,981.04
          Travel      gas       $1,200.00     $984.93     $215.07
                      leases    $4,500.00    $3,889.65     $610.35
```

Example 3: Grouping and Summarizing Observations

This example groups rows of the report by DEPT and ACCOUNT. It also summarizes the data for each department and for the entire company. The report appears in Output 10.6. The REPORT language that creates this report follows:

```
libname report 'SAS-data-library';

proc report data=report.budget;          1

   title;

   column ('Year-to-Date Financial Status'      2
           'Grouped by Department' ' '
           dept account budget actual balance);

   define dept    / group format=$10. width=10 'Department';   3
   define account / group format=$8. width=8 'Account';
   define budget  / sum format=dollar11.2 width=11
                    'Amount/Budgeted';
   define actual  / sum format=dollar11.2 width=11
                    'Amount/Spent';
   define balance / computed format=dollar11.2 width=11
                    'Balance';
```

```
compute balance;        4
   balance=budget.sum-actual.sum;
endcomp;

break after dept / ol ul skip summarize suppress color=red;     5

rbreak after / dol dul summarize color=red;     6
run;
```

1. Invoke the REPORT procedure.

2. Create a three-line header that spans all columns of the report. The first two lines contain text; the third is blank. (Because of the null TITLE statement, this header appears to be a title.) Create one column for each variable in the data set except QTR and one column for the computed variable BALANCE.

3. Define the data set variables and the computed variable, BALANCE.

4. Assign a value to the computed variable, BALANCE. Because you are calculating BALANCE from analysis variables, you must use compound names in the assignment statement.

5. Create a break in the report each time the value of DEPT changes. The break occurs after the last row in each set of rows that have the same value of DEPT. The break contains overlining, underlining, summaries of the statistics and computed variable, and a blank line as the last break line.
    Note: Currently, color appears only when PROC REPORT displays the report in the REPORT window.

6. Create a break at the end of the report. The break contains double overlining, double underlining, and summaries of the statistics and computed variable.

Output 10.6
Grouping and
Summarizing
Observations

```
                                                              1
                   Year-to-Date Financial Status
                     Grouped by Department

                          Amount       Amount
     Department  Account   Budgeted      Spent       Balance
     Equipment   lease    $80,000.00   $80,000.00        $0.00
                 maint    $22,000.00   $18,217.42    $3,782.58
                 purchase $60,000.00   $66,051.53   $-6,051.53
                 rental   $10,000.00    $9,481.81      $518.19
                 sets     $15,000.00   $16,422.30   $-1,422.30
                 tape     $20,000.00   $18,256.15    $1,743.85
                          -----------  -----------  -----------
                         $207,000.00  $208,429.21   $-1,429.21
                          -----------  -----------  -----------

     Facilities  rent     $48,000.00   $48,000.00        $0.00
                 supplies  $5,500.00    $4,959.03      $540.97
                 utils     $8,500.00    $7,668.10      $831.90
                          -----------  -----------  -----------
                          $62,000.00   $60,627.13    $1,372.87
                          -----------  -----------  -----------

     Other       advert   $60,000.00   $69,802.62   $-9,802.62
                 musicfee  $8,000.00    $7,426.45      $573.55
                 talent   $33,000.00   $31,411.37    $1,588.63
                          -----------  -----------  -----------
                         $101,000.00  $108,640.44   $-7,640.44
                          -----------  -----------  -----------
```

```
      Staff      fulltime  $295,000.00  $293,988.43   $1,011.57
                 parttime  $100,000.00   $99,869.08     $130.92
                           -----------  -----------  -----------
                           $395,000.00  $393,857.51   $1,142.49
                           -----------  -----------  -----------

      Travel     gas         $2,000.00    $1,522.19     $477.81
                 leases      $8,000.00    $6,934.80   $1,065.20
                           -----------  -----------  -----------
                            $10,000.00    $8,456.99   $1,543.01
                           -----------  -----------  -----------


                           ===========  ===========  ===========
                           $775,000.00  $780,011.28  $-5,011.28
                           ===========  ===========  ===========
```

Example 4: Using the Same Variable in Multiple Ways

This example uses QTR as an across variable, creating one column in the report for each value of QTR. It also uses the variable BUDGET in two different ways: once sharing a column with QTR and once in a column by itself. The report appears in Output 10.7. The REPORT language that creates this report follows.

Note: You do not need an alias for BUDGET even though it appears twice in the COLUMN statement because its characteristics are the same in both places you use it.

```
libname report 'SAS-data-library';

proc format;
   value forqtr 1='1st' 2='2nd';          1
run;

proc report data=report.budget;          2

   column dept account qtr,budget budget;          3

   define dept    / group format=$10. width=10 'DEPT';          4
   define account / group format=$8. width=8 'ACCOUNT';

   define qtr     / across format=forqtr12. width=12 '_QTR_';          5
   define budget  / sum format=dollar11.2 width=11 'BUDGET';          6

run;
```

1. Define the FORQTR format.

2. Invoke the REPORT procedure.

3. Create a column for DEPT, ACCOUNT, QTR, BUDGET below QTR, and BUDGET to the right of QTR.

4. Define the variables DEPT and ACCOUNT as in previous examples.

5. Define QTR as an across variable. Assign a format of FORQTR12., a width of 12, and a column header of '_QTR_'.

6. Define BUDGET as an analysis variable used to calculate the SUM statistic. Assign a format of DOLLAR11.2, a width of 11, and a column header of 'BUDGET'.

Output 10.7
Using a Variable
in Multiple Ways

```
                              The SAS System                            1

                                 ___QTR___
                                  1st          2nd
           DEPT       ACCOUNT    BUDGET       BUDGET        BUDGET
           Equipment  lease     $40,000.00   $40,000.00    $80,000.00
                      maint     $10,000.00   $12,000.00    $22,000.00
                      purchase  $40,000.00   $20,000.00    $60,000.00
                      rental     $4,000.00    $6,000.00    $10,000.00
                      sets       $7,500.00    $7,500.00    $15,000.00
                      tape       $8,000.00   $12,000.00    $20,000.00
           Facilities rent      $24,000.00   $24,000.00    $48,000.00
                      supplies   $2,750.00    $2,750.00     $5,500.00
                      utils      $5,000.00    $3,500.00     $8,500.00
           Other      advert    $30,000.00   $30,000.00    $60,000.00
                      musicfee   $3,000.00    $5,000.00     $8,000.00
                      talent    $13,500.00   $19,500.00    $33,000.00
           Staff      fulltime $130,000.00  $165,000.00   $295,000.00
                      parttime  $40,000.00   $60,000.00   $100,000.00
           Travel     gas          $800.00    $1,200.00     $2,000.00
                      leases     $3,500.00    $4,500.00     $8,000.00
```

Example 5: Creating a Sophisticated Report

This report uses QTR as an across variable, displaying values for the balance in each account in each department for each quarter and for the year to date. A customized break line appears each time the value of DEPT changes. The break line tells you how much of the company's year-to-date budget was allocated to the department. The report appears in Output 10.8. The REPORT language that creates this report follows:

```
libname report 'SAS-data-library';
options linesize=255 nocenter;         1

proc format;         2
   value forqtr 1='1st' 2='2nd';
run;

proc report data=report.budget;         3
   title;

   column dept account qtr,(budget actual qtrbal)    4
          budget actual yrtodate z;
```

```
   define dept      / group format=$10. width=10 'Department';   5
   define account   / group format=$8. width=8 'Account';
   define qtr       / across format=forqtr12. width=12 '_Quarter_';
   define qtrbal    / computed format=dollar11.2 width=11 'Balance';
   define budget    / sum noprint;
   define actual    / sum noprint;
   define yrtodate  / computed format=dollar11.2 width=12
                      'Year-to-Date/Balance';
   define  z        / computed format=$1. width=1 spacing=0 ' ';

   compute qtrbal;        6
      _c5_=_c3_-_c4_;
      _c8_=_c6_-_c7_;
   endcomp;

   compute yrtodate;        7
      yrtodate=budget.sum-actual.sum;
   endcomp;

   compute z / length=1 char;
      if yrtodate<0 then z='*';        8
      else z=' ';
   endcomp;

   break after dept / ol page summarize color=red;        9

   compute after dept;        10
      pctbud=round((budget.sum/totalbud)*100);
      line ' ';
      line pctbud 2. '% of the year-to-date budget is allocated to this department.';
   endcomp;

   compute before ;        11
      totalbud=budget.sum;
   endcomp;
run;
```

1. Set the appropriate options. The line size of 255 is to accommodate the SHOWALL option, which you may want to use to see the columns suppressed by the NOPRINT option.

2. Define the QTR format.

3. Invoke the REPORT procedure.

4. Create a column for DEPT, ACCOUNT, QTR, BUDGET, ACTUAL, YRTODATE, and Z. Below the column for QTR create columns for BUDGET, ACTUAL, and QTRBAL.

5. Set the characteristics for the variables in the report. Note that you don't need to set any characteristics besides usage for variables for which you specify the NOPRINT option.

6. Compute the values for QTRBAL based on column numbers.

7. Compute the value of YRTODATE based on compound names.

8. Compute the value of Z.

9. Create a break in the report each time the value of DEPT changes. The break occurs after the last row in each set of rows that have the same value of DEPT. The break contains overlining, summaries of the statistics and computed variable, and a page break. It does not contain the value of the break variable. All break lines are red.

   Note: Currently, color appears only when PROC REPORT displays the report in the REPORT window.

10. Create a customized break line at the break after DEPT.

11. At a break at the beginning of the report, save the value of the total budget for the year to date.

Output 10.8
Creating a
Sophisticated
Report

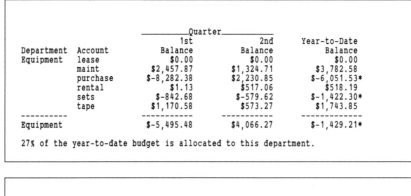

```
                              _____Quarter_____
                                  1st            2nd        Year-to-Date
Department  Account            Balance        Balance         Balance
Equipment   lease                $0.00          $0.00           $0.00
            maint            $2,457.87      $1,324.71       $3,782.58
            purchase        $-8,282.38      $2,230.85      $-6,051.53*
            rental               $1.13        $517.06         $518.19
            sets              $-842.68       $-579.62      $-1,422.30*
            tape             $1,170.58        $573.27       $1,743.85
----------                  -----------    -----------    -------------
Equipment                   $-5,495.48      $4,066.27      $-1,429.21*

27% of the year-to-date budget is allocated to this department.
```

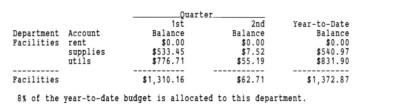

```
                              _____Quarter_____
                                  1st            2nd        Year-to-Date
Department  Account            Balance        Balance         Balance
Facilities  rent                 $0.00          $0.00           $0.00
            supplies          $533.45          $7.52         $540.97
            utils             $776.71         $55.19         $831.90
----------                  -----------    -----------    -------------
Facilities                  $1,310.16         $62.71       $1,372.87

8% of the year-to-date budget is allocated to this department.
```

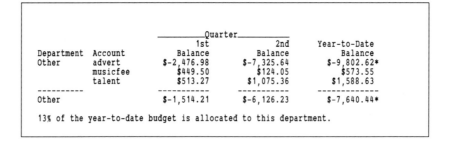

```
                              _____Quarter_____
                                  1st            2nd        Year-to-Date
Department  Account            Balance        Balance         Balance
Other       advert         $-2,476.98      $-7,325.64      $-9,802.62*
            musicfee          $449.50        $124.05         $573.55
            talent            $513.27      $1,075.36       $1,588.63
----------                  -----------    -----------    -------------
Other                       $-1,514.21     $-6,126.23      $-7,640.44*

13% of the year-to-date budget is allocated to this department.
```

```
                           _____Quarter_____
                           1st          2nd        Year-to-Date
Department  Account       Balance      Balance       Balance
Staff       fulltime      $2,357.32   $-1,345.75    $1,011.57
            parttime      $-3,850.12   $3,981.04     $130.92
----------               -----------  -----------   ------------
Staff                     $-1,492.80   $2,635.29    $1,142.49

51% of the year-to-date budget is allocated to this department.
```

```
                           _____Quarter_____
                           1st          2nd        Year-to-Date
Department  Account       Balance      Balance       Balance
Travel      gas           $262.74      $215.07       $477.81
            leases        $454.85      $610.35      $1,065.20
----------               -----------  -----------   ------------
Travel                    $717.59      $825.42      $1,543.01

1% of the year-to-date budget is allocated to this department.
```

Example 6:   Creating a Report with Multiple Code Segments

This report uses code segments attached to a variety of places in the report to produce a highly customized report. The report appears in Output 10.9. The REPORT language that creates this report follows:

```
proc format;      1
   value forqtr 1='1st' 2='2nd';
run;

title;

proc report data=report.budget;      2

   column qtr dept budget actual balance budget=bud2 actual=act2; 3

   define qtr     / group format=3. width=12 'Quarter';      4
   define dept    / group format=$10. width=10 'Department';
   define budget  / sum noprint;
   define actual  / sum noprint;
   define balance / computed format=dollar11.2 width=11 'Balance';
   define bud2    / sum format=dollar11.2 width=11 'Budget';
   define act2    / sum format=dollar11.2 width=11 'Actual';
```

```
compute balance;
   balance=budget.sum-actual.sum;        5
   if dept ne ' ' then                6
      do;
         if balance<minbal then
            do;
               minbal=balance;
               mindept=dept;
               minqtr=holdqtr;
            end;
         if balance<qminbal then
            do;
               qminbal=balance;
               qmindept=dept;
            end;
      end;
endcomp;


compute before qtr;        7
   holdqtr=qtr;
   qminbal=0;
   qmindept=' ';
endcomp;


break after qtr / color=red;        8

compute after qtr;
   length text3 text4 $ 75;        9
   if qminbal=0 then
      do;
         text3='No departments were over budget this quarter.';
         text4=' ';
      end;
   else
      do;
         text3='The largest overdraw for this quarter was in the '
               || trim(qmindept) || ' department.';
         text4='It was overdrawn by '
               || left(put(abs(qminbal),dollar11.2)) || '.';
      end;
   line ' ';        10
   line text3 $75.;
   line text4 $75.;
   line ' ';
endcomp;

compute before ;        11
   minbal=0;
   mindept=' ';
endcomp;

rbreak after  / ol ul summarize color=red;        12
```

```
   compute after ;
      dept='Total:';
      length text1 text2 $ 75;        13
      if minbal=0 then
         do;
            text1='No departments were over budget.';
            text2=' ';
         end;
      else
         do;
            text1='The largest overdraw was in the ' || trim(mindept)
                  || ' department during the '
                  || put(minqtr,forqtr3.) || ' quarter.';
            text2='It was overdrawn by '
                  || left(put(abs(minbal),dollar11.2)) || '.';
         end;
      line ' ';         14
      line 75*'=';
      line text1 $75.;
      line text2 $75.;
      line 75*'=';
   endcomp;

run;
```

1. Create the format FORQTR. to use in the report.

2. Invoke the REPORT procedure.

3. Create a column for QTR, DEPT, BUDGET, ACTUAL, BALANCE, BUDGET (with the alias BUD2), and ACTUAL (with the alias ACT2).

4. Set the characteristics for the variables in the report. Note that you don't need to set any characteristics besides usage for variables for which you specify the NOPRINT option. Also, notice the definitions for the variables with aliases (BUD2 and ACT2).

5. Compute the value of BALANCE based on compound names.

6. On each detail row, compare the current value of BALANCE to the minimum for the year and to the minimum for the quarter. Store the values in the current row if the value of BALANCE is a new minimum.

7. At a break before QTR, initialize values for HOLDQTR, QMINBAL, and QMINDEPT.

8. Create a break after QTR. Set the color of all break lines to red.

9. At the break after QTR, create two text strings to use in the break lines. The contents of the text strings depend on whether or not any departments are overbudget during the current quarter.

10. Write the break lines for the break after QTR.

11. At a break at the beginning of the report, initialize MINBAL and MINDEPT.

12. Create a break at the end of the report. This break includes overlining, underlining, and a summary of the statistics and computed variable. All break lines are red.

13. At the break at the end of the report, create two text strings to use in the break lines. The contents of the text strings depend on whether or not any departments are overbudget for the year to date.

14. Write the break lines at the end of the report.

    Note: Currently, color appears only when PROC REPORT displays the report in the REPORT window.

Output 10.9
Creating a Report with Multiple Code Segments

```
                                                                                    1
         Quarter  Department         Balance       Budget       Actual
               1  Equipment        $-5,495.48  $109,500.00  $114,995.48
                  Facilities        $1,310.16   $31,750.00   $30,439.84
                  Other            $-1,514.21   $46,500.00   $48,014.21
                  Staff            $-1,492.80  $170,000.00  $171,492.80
                  Travel             $717.59    $4,300.00    $3,582.41
   The largest overdraw for this quarter was in the Equipment department.
   It was overdrawn by $5,495.48.

               2  Equipment        $4,066.27   $97,500.00   $93,433.73
                  Facilities         $62.71    $30,250.00   $30,187.29
                  Other            $-6,126.23   $54,500.00   $60,626.23
                  Staff            $2,635.29   $225,000.00  $222,364.71
                  Travel            $825.42     $5,700.00    $4,874.58
   The largest overdraw for this quarter was in the Other department.
   It was overdrawn by $6,126.23.

                  ----------    -----------  -----------  -----------
                  Total:        $-5,011.28   $775,000.00  $780,011.28
                  ----------    -----------  -----------  -----------


   =====================================================================
   The largest overdraw was in the Other department during the 2nd quarter.
   It was overdrawn by $6,126.23.
   =====================================================================
```

Part 3
Appendix

Appendix   **Creating the Data Set REPORT.BUDGET**

Appendix

Creating the Data Set REPORT.BUDGET

This appendix contains the DATA step and the raw data file that create the data set REPORT.BUDGET, which all the examples in this book use. In order to store the data set permanently, you must assign a libref to a permanent SAS data library and use that libref in the DATA statement. The following SAS statements store the data set in the library REPORT:

```
libname report 'SAS-data-library';

data report.budget;
   infile 'your-input-file';
   input qtr a3 dept $ 3-12 a14 account $ a23 budget a30 actual;
   actual=actual/100;
   format budget actual dollar11.2;
run;
```

These statements read raw data from the external file you specify as *your-input-file*. The external file appears below:

```
----+----1----+----2----+----3----+----4----+----5----+----6----+----7--
1 Staff      fulltime 130000 12764268
2 Staff      fulltime 165000 16634575
1 Staff      parttime 40000  4385012
2 Staff      parttime 60000  5601896
1 Equipment  lease    40000  4000000
2 Equipment  lease    40000  4000000
1 Equipment  purchase 40000  4828238
2 Equipment  purchase 20000  1776915
1 Equipment  tape     8000   682942
2 Equipment  tape     12000  1142673
1 Equipment  sets     7500   834268
2 Equipment  sets     7500   807962
1 Equipment  maint    10000  754213
2 Equipment  maint    12000  1067529
1 Equipment  rental   4000   399887
2 Equipment  rental   6000   548294
1 Facilities rent     24000  2400000
2 Facilities rent     24000  2400000
1 Facilities utils    5000   422329
2 Facilities utils    3500   344481
1 Facilities supplies 2750   221655
2 Facilities supplies 2750   274248
1 Travel     leases   3500   304515
2 Travel     leases   4500   388965
1 Travel     gas      800    53726
2 Travel     gas      1200   98493
1 Other      advert   30000  3247698
2 Other      advert   30000  3732564
----+----1----+----2----+----3----+----4----+----5----+----6----+----7--
```

(continued on next page)

(continued from previous page)

```
----+----1----+----2----+----3----+----4----+----5----+----6----+----7--
1 Other      talent   13500  1298673
2 Other      talent   19500  1842464
1 Other      musicfee 3000   255050
2 Other      musicfee 5000   487595
```

Glossary

across variable

a variable used so that each formatted value of the variable forms a column in the report.

action bar

a list of selections across the top of a window. The action bars used in REPORT windows contain one or more of the following selections: File, Edit, Search, View, Locals, and Globals. When you make a selection from the action bar, the REPORT procedure displays a pull-down menu. You can create your own action bars for the REPORT and COMPUTE windows with the PMENU procedure. To use these action bars, point to them in the PROFILE window of the REPORT procedure.

alias

an alternate name for a report item. An alias enables you to set different characteristics for the same item when you use the item in more than one way in the report.

analysis variable

a variable for which you can calculate statistics. You must associate a statistic with an analysis variable. You specify the statistic in the Statistic field of the DEFINITION window or with an option in the DEFINE statement. In conjunction with group variables, an analysis variable can consolidate multiple observations that have a unique combination of values for all group variables into one row. The value displayed for an analysis variable is the statistic associated with it calculated for the set of observations represented by that row and column of the report.

By default, the REPORT procedure treats a numeric variable as an analysis variable used to calculate the SUM statistic.

background

background of a window.

banner

the command prompt in the upper-left corner of a window.

border

the line drawn around the window, including the window name.

break

a section of the report that does one or more of the following: visually separates parts of the report; summarizes statistics and computed variables; displays text, values calculated for a set of rows of the report, or both; executes DATA step statements. You can create breaks when the value of a selected variable changes or at the beginning or end of a report. See also break variable.

break line

a line of a report that contains one of the following: characters that visually separate parts of the report; summaries of statistics and computed variables (called a summary line); text, values calculated for a set of rows of the report, or both.

break variable

a group or order variable that you select to determine the location of break lines. The REPORT procedure performs the actions you specify for the break each time the value of this variable changes.

check box

in a window, one choice in a list of choices in which each item is preceded by a small box. You can select multiple check boxes from one list.

code segment

the statements that appear in one COMPUTE window or between one pair of COMPUTE and ENDCOMP statements.

command

in a window, the action bar or the unprotected part of the command line where you enter your commands.

compound name

a name that has the following form:

variable-name.statistic

A compound name is used to refer to the value of a statistic.

computed variable

a variable whose value you calculate based on other variables in the report. A computed variable becomes part of the report, but it is not part of the input data set.

data set variable

a variable in the input data set.

DATA step variable

a variable that appears in one or more code segments but does not appear in a column of the report.

detail row

a row of a report that either contains information from a single observation in the data set or consolidates the information for a group of observations that have a unique combination of values for all group variables.

display variable

a variable that does not affect the order of the rows of the report. A report that contains one or more display variables has a detail row for each observation in the data set.

By default, the REPORT procedure treats character variables as display variables.

field

an area in a window in which you can enter or edit a value.

group variable

a variable that orders the detail rows in a report according to their formatted values and consolidates multiple observations that have a unique combination of values for all group variables into one row.

header

a string of characters that spans one or more columns in the report. A header can occupy multiple lines. See also split character.

item

a data set variable, a statistic, or a computed variable. In many cases an item occupies a single column of the report, but an item can occupy multiple columns. Under some circumstances, multiple items can share a column.

message

the contents of the line immediately below the command line or action bar in a window. This line is known as the message line.

order variable

a variable that orders the detail rows in a report according to their formatted values. A report that contains one or more order variables has a detail row for every observation in the data set.

panel

a set of columns in a report. A familiar example of a report with panels is a telephone book, which contains multiple panels of names and telephone numbers on a single page.

PROMPT facility

a tool, implemented through the PROMPTER window, that prompts you for information as you add either data set variables or statistics to a report. It steps you through the most commonly used parts of the windows you would use to add either a data set variable or a statistic to a report. The PROMPTER window provides more guidance than the other windows provide.

pull-down menu

a list of REPORT commands, global SAS commands frequently used with the REPORT procedure, or both, that appears when you make a selection from an action bar.

push button

a part of a window that is always highlighted. The selection of a push button initiates an action, such as exiting a window.

radio box

a list in a window from which you can select only one item. Each element in the list is preceded by an open circle.

radio button

an item in a radio box.

report break

a break at the beginning or end of a report. See also break.

report variable

a variable that constitutes one or more columns of the report. The variable may or may not appear in one or more code segments.

Note: You can suppress the display of a report variable with the NOPRINT or the NOZERO option in the DEFINITION window or in the DEFINE statement.

SAS data library

a collection of one or more SAS files that are recognized by the SAS System. Each file is a member of the library.

split character

a character that splits headers across multiple lines. If you use the split character in a column header, the REPORT procedure breaks the header when it reaches that character and continues the header on the next line. The split character itself is not part of the column header.

summary line

a break line that summarizes statistics and computed variables.

Index

Special Characters

Your Turn

If you have comments or suggestions about the *SAS Guide to the REPORT Procedure: Usage and Reference, Version 6, First Edition*, or the REPORT procedure, please send them to us on a photocopy of this page.

Please send the photocopy to the Publications Division (for comments about this book) or the Technical Support Division (for suggestions about the REPORT procedure) at SAS Institute Inc., SAS Campus Drive, Cary, NC 27513.